Learn

Eureka Math®

Grade 3
Modules 5 & 6

Published by Great Minds®.

Printed in the U.S.A.
This book may be purchased from the publisher at eureka-math.org.
1 2 3 4 5 6 7 8 9 10 CCR 24 23 22

ISBN 978-1-64054-062-0

G3-M5-M6-L-05.2018

Learn • Practice • Succeed

Eureka Math® student materials for *A Story of Units*® (K–5) are available in the *Learn, Practice, Succeed* trio. This series supports differentiation and remediation while keeping student materials organized and accessible. Educators will find that the *Learn, Practice,* and *Succeed* series also offers coherent—and therefore, more effective—resources for Response to Intervention (RTI), extra practice, and summer learning.

Learn

Eureka Math Learn serves as a student's in-class companion where they show their thinking, share what they know, and watch their knowledge build every day. *Learn* assembles the daily classwork—Application Problems, Exit Tickets, Problem Sets, templates—in an easily stored and navigated volume.

Practice

Each *Eureka Math* lesson begins with a series of energetic, joyous fluency activities, including those found in *Eureka Math Practice.* Students who are fluent in their math facts can master more material more deeply. With *Practice,* students build competence in newly acquired skills and reinforce previous learning in preparation for the next lesson.

Together, *Learn* and *Practice* provide all the print materials students will use for their core math instruction.

Succeed

Eureka Math Succeed enables students to work individually toward mastery. These additional problem sets align lesson by lesson with classroom instruction, making them ideal for use as homework or extra practice. Each problem set is accompanied by a Homework Helper, a set of worked examples that illustrate how to solve similar problems.

Teachers and tutors can use *Succeed* books from prior grade levels as curriculum-consistent tools for filling gaps in foundational knowledge. Students will thrive and progress more quickly as familiar models facilitate connections to their current grade-level content.

Students, families, and educators:

Thank you for being part of the *Eureka Math®* community, where we celebrate the joy, wonder, and thrill of mathematics.

In the *Eureka Math* classroom, new learning is activated through rich experiences and dialogue. The *Learn* book puts in each student's hands the prompts and problem sequences they need to express and consolidate their learning in class.

What is in the Learn book?

Application Problems: Problem solving in a real-world context is a daily part of *Eureka Math.* Students build confidence and perseverance as they apply their knowledge in new and varied situations. The curriculum encourages students to use the RDW process—Read the problem, Draw to make sense of the problem, and Write an equation and a solution. Teachers facilitate as students share their work and explain their solution strategies to one another.

Problem Sets: A carefully sequenced Problem Set provides an in-class opportunity for independent work, with multiple entry points for differentiation. Teachers can use the Preparation and Customization process to select "Must Do" problems for each student. Some students will complete more problems than others; what is important is that all students have a 10-minute period to immediately exercise what they've learned, with light support from their teacher.

Students bring the Problem Set with them to the culminating point of each lesson: the Student Debrief. Here, students reflect with their peers and their teacher, articulating and consolidating what they wondered, noticed, and learned that day.

Exit Tickets: Students show their teacher what they know through their work on the daily Exit Ticket. This check for understanding provides the teacher with valuable real-time evidence of the efficacy of that day's instruction, giving critical insight into where to focus next.

Templates: From time to time, the Application Problem, Problem Set, or other classroom activity requires that students have their own copy of a picture, reusable model, or data set. Each of these templates is provided with the first lesson that requires it.

Where can I learn more about Eureka Math *resources?*

The Great Minds® team is committed to supporting students, families, and educators with an ever-growing library of resources, available at eureka-math.org. The website also offers inspiring stories of success in the *Eureka Math* community. Share your insights and accomplishments with fellow users by becoming a *Eureka Math* Champion.

Best wishes for a year filled with aha moments!

Jill Diniz
Director of Mathematics
Great Minds

The Read–Draw–Write Process

The *Eureka Math* curriculum supports students as they problem-solve by using a simple, repeatable process introduced by the teacher. The Read–Draw–Write (RDW) process calls for students to

1. Read the problem.
2. Draw and label.
3. Write an equation.
4. Write a word sentence (statement).

Educators are encouraged to scaffold the process by interjecting questions such as

- What do you see?
- Can you draw something?
- What conclusions can you make from your drawing?

The more students participate in reasoning through problems with this systematic, open approach, the more they internalize the thought process and apply it instinctively for years to come.

Contents

Module 5: Fractions as Numbers on the Number Line

Module 6: Collecting and Displaying Data

Grade 3
Module 5

Measure the length of your math book using a ruler in inches. Then measure it again in centimeters.

a. Which is a larger unit, an inch or a centimeter?

b. Which would yield a greater number when measuring the math book, inches or centimeters?

Read **Draw** **Write**

c. Measure at least 2 different items in both inches and centimeters. What do you notice?

Read **Draw** **Write**

Name ______________________________ Date ______________

1. A beaker is considered full when the liquid reaches the fill line shown near the top. Estimate the amount of water in the beaker by shading the drawing as indicated. The first one is done for you.

1 half

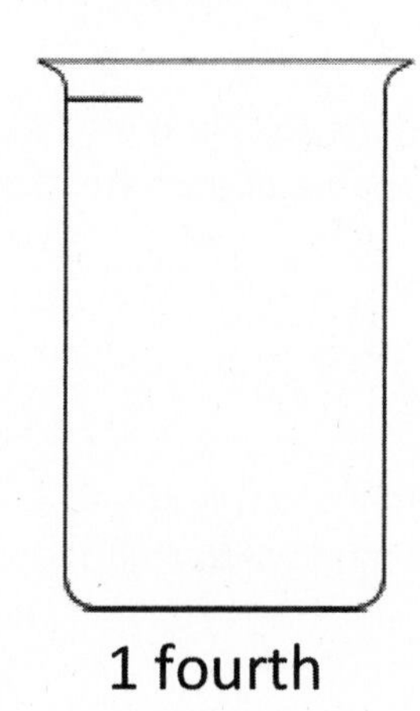
1 fourth

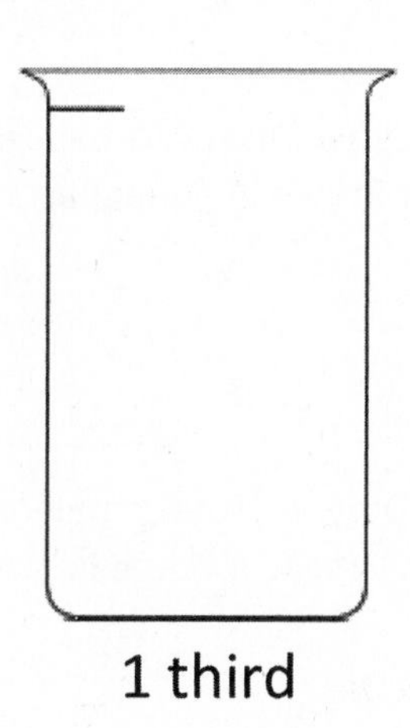
1 third

2. Juanita cut her string cheese into equal pieces as shown in the rectangles below. In the blanks below, name the fraction of the string cheese represented by the shaded part.

____

____

____

3. a. In the space below, draw a small rectangle. Estimate to split it into 2 equal parts. How many lines did you draw to make 2 equal parts? What is the name of each fractional unit?

   b. Draw another small rectangle. Estimate to split it into 3 equal parts. How many lines did you draw to make 3 equal parts? What is the name of each fractional unit?

   c. Draw another small rectangle. Estimate to split it into 4 equal parts. How many lines did you draw to make 4 equal parts? What is the name of each fractional unit?

4. Each rectangle represents 1 sheet of paper.

   a. Estimate to show how you would cut the paper into fractional units as indicated below.

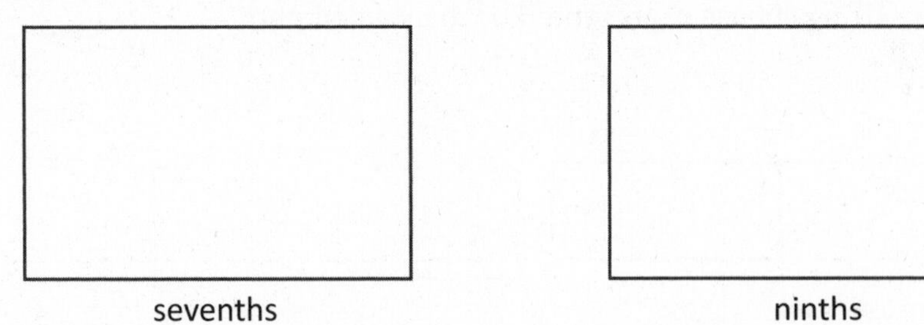

seventhsninths

   b. What do you notice? How many lines do you think you would draw to make a rectangle with 20 equal parts?

5. Rochelle has a strip of wood 12 inches long. She cuts it into pieces that are each 6 inches in length. What fraction of the wood is one piece? Use your strip from the lesson to help you. Draw a picture to show the piece of wood and how Rochelle cut it.

Name ____ Date ____________________

1. Name the fraction that is shaded.

____

2. Estimate to partition the rectangle into thirds.

3. A plumber has 12 feet of pipe. He cuts it into pieces that are each 3 feet in length. What fraction of the pipe would one piece represent? (Use your strip from the lesson to help you.)

Anu needs to cut a piece of paper into 6 equal parts. Draw at least 3 pictures to show how Anu can cut her paper so that all the parts are equal.

Read **Draw** **Write**

Name ______________________________ Date ______________

1. Circle the strips that are folded to make equal parts.

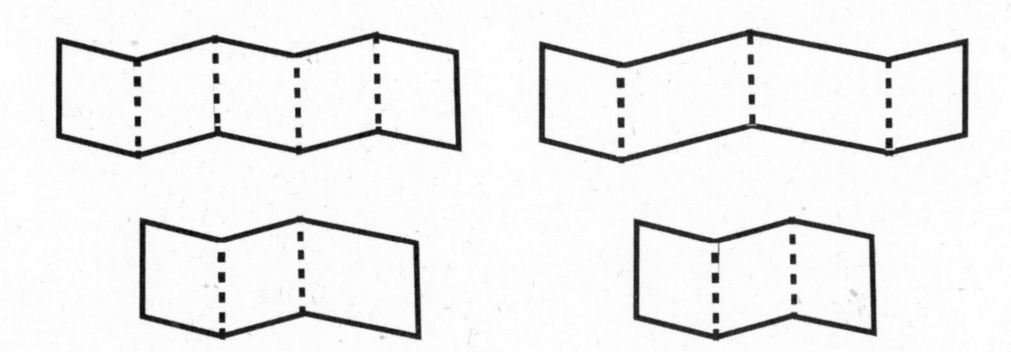

2.

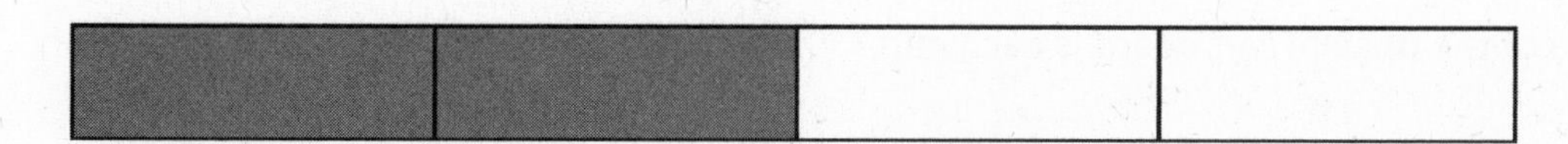

a. There are ______ equal parts in all. ______ are shaded.

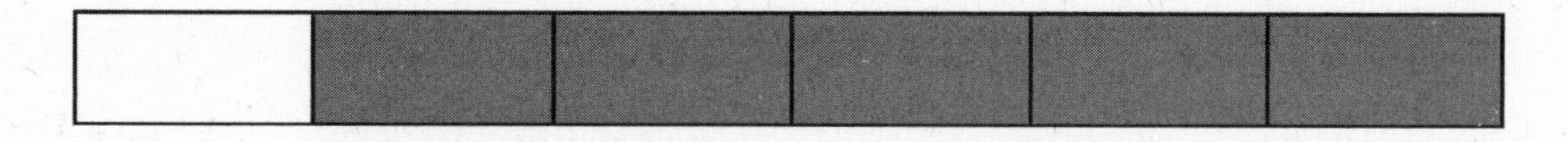

b. There are ______ equal parts in all. ______ are shaded.

c. There are ______ equal parts in all. ______ are shaded.

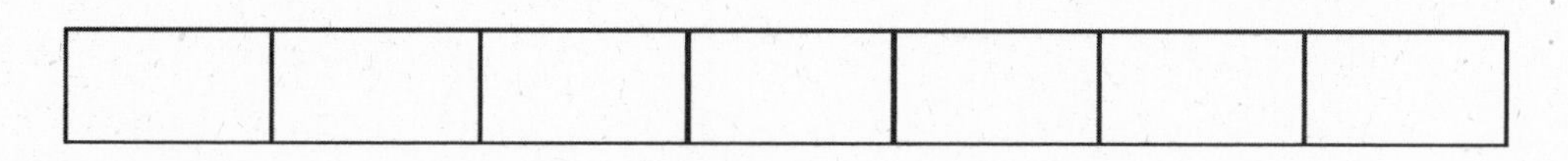

d. There are ______ equal parts in all. ______ are shaded.

Use your fraction strips as tools to help you solve the following problems.

3. Noah, Pedro, and Sharon share a whole candy bar fairly. Which of your fraction strips shows how they each get an equal part? Draw the candy bar below. Then, label Sharon's fraction of the candy bar.

4. To make a garage for his toy truck, Zeno bends a rectangular piece of cardboard in half. He then bends each half in half again. Which of your fraction strips best matches this story?

   a. What fraction of the original cardboard is each part? Draw and label the matching fraction strip below.

   b. Zeno bends a different piece of cardboard in thirds. He then bends each third in half again. Which of your fraction strips best matches this story? Draw and label the matching fraction strip in the space below.

Name ______________________________ Date ______________

1. Circle the model that correctly shows 1 third shaded.

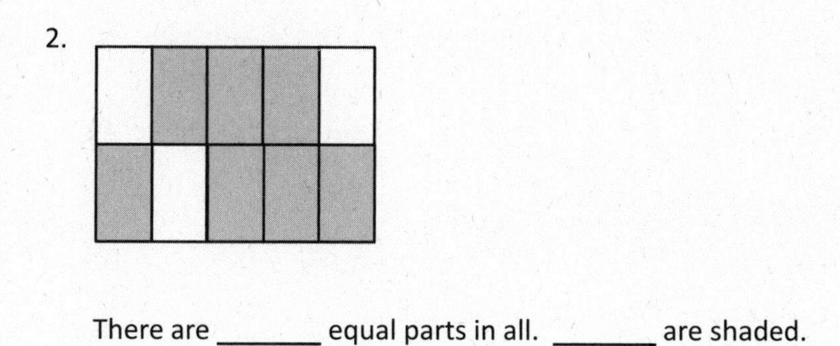

2.

There are ______ equal parts in all. ______ are shaded.

3. Michael bakes a piece of garlic bread for dinner. He shares it equally with his 3 sisters. Show how Michael and his 3 sisters can each get an equal share of the garlic bread.

Marcos has a 1-liter jar of milk to share with his mother, father, and sister. Draw a picture to show how Marcos must share the milk so that everyone gets the same amount. What fraction of the milk does each person get?

Read **Draw** **Write**

Name ______________________________ Date ______________

1. Each shape is a whole divided into equal parts. Name the fractional unit, and then count and tell how many of those units are shaded. The first one is done for you.

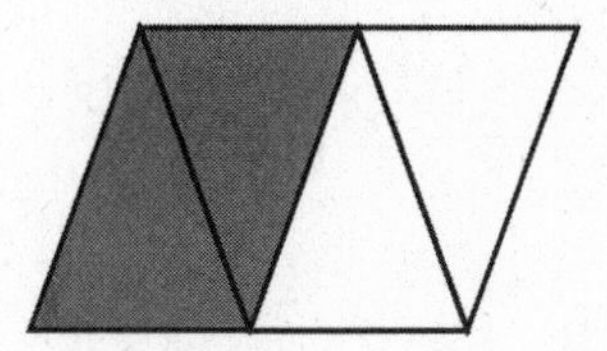

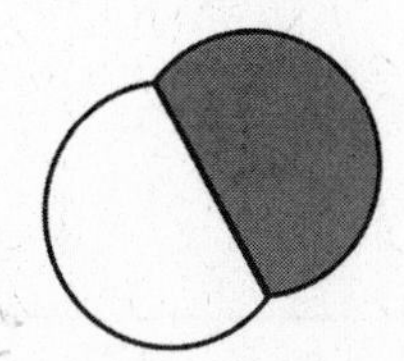

Fourths
2 fourths are shaded.

2. Circle the shapes that are divided into equal parts. Write a sentence telling what *equal parts* means.

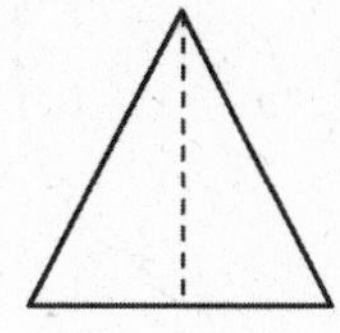
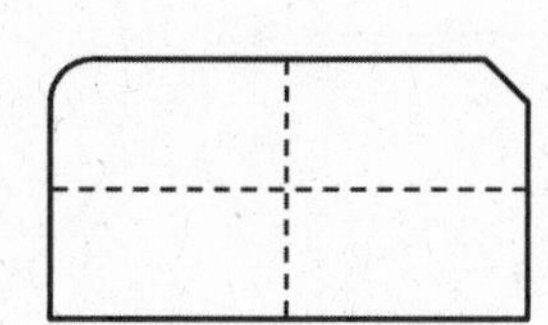
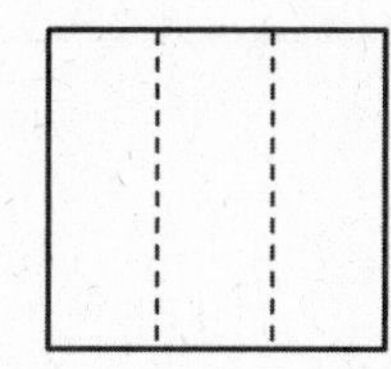
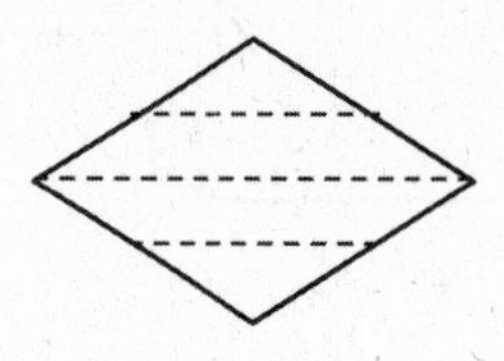
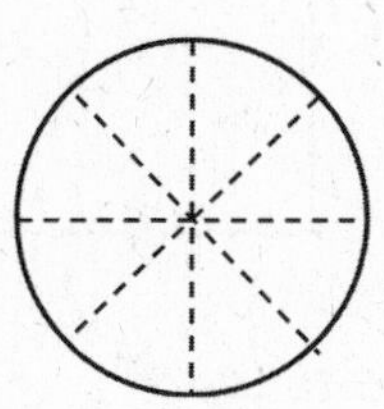

3. Each shape is 1 whole. Estimate to divide each into 4 equal parts. Name the fractional unit below.

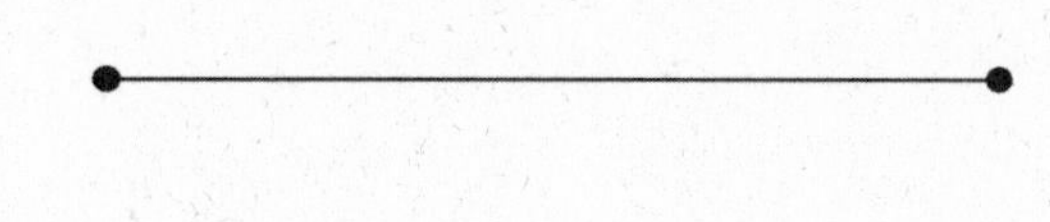
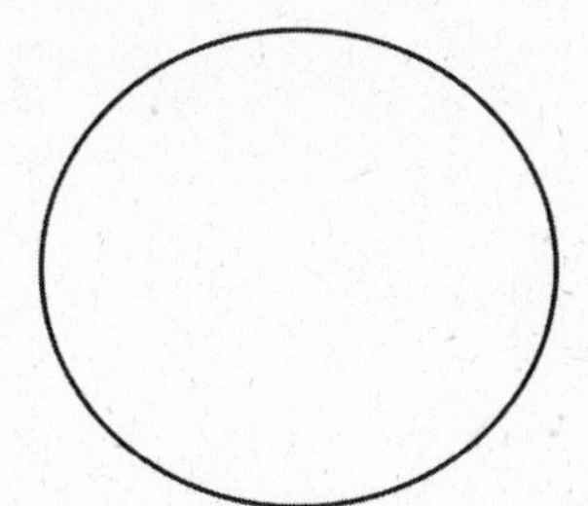

Fractional unit: ______________

4. Each shape is 1 whole. Divide and shade to show the given fraction.

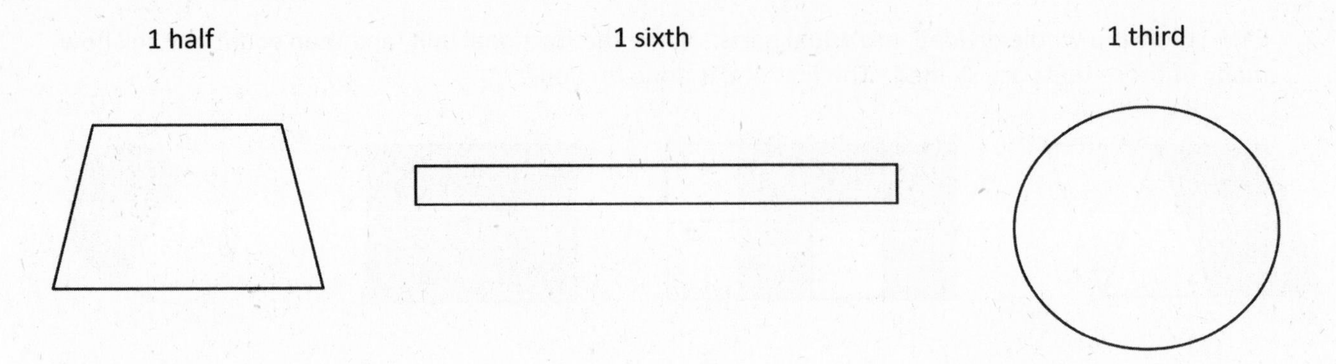

5. Each shape is 1 whole. Estimate to divide each into equal parts (do not draw fourths). Divide each whole using a different fractional unit. Write the name of the fractional unit on the line below the shape.

______________ ______________ ______________

6. Charlotte wants to equally share a candy bar with 4 friends. Draw Charlotte's candy bar. Show how she can divide her candy bar so everyone gets an equal share. What fraction of the candy bar does each person receive?

Each person receives ____________________.

Name ______________________________ Date ______________

1. 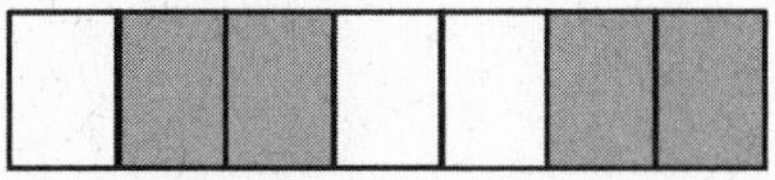 [bar with 7 parts] ____________ sevenths are shaded.

2. Circle the shapes that are divided into equal parts.

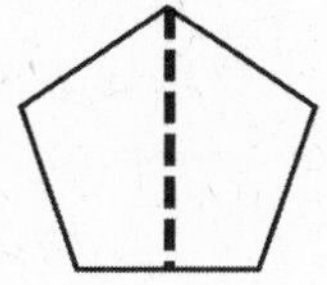
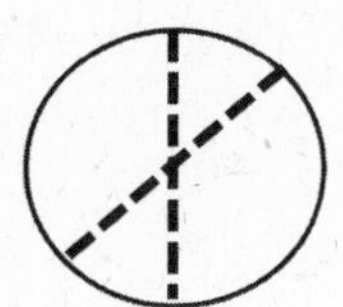
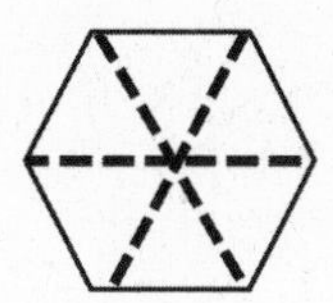

3. Steven wants to equally share his pizza with his 3 sisters. What fraction of the pizza does he and each sister receive?

He and each sister receive ______________________

Mr. Ramos sliced an orange into 8 equal pieces. He ate 1 slice. Draw a picture to represent the 8 slices of an orange. Shade in the slice Mr. Ramos ate. What fraction of the orange did Mr. Ramos eat? What fraction did he not eat?

Read Draw Write

Name ______________________________ Date ____________________

1. Draw a picture of the yellow strip at 3 (or 4) different stations. Shade and label 1 fractional unit of each.

2. Draw a picture of the brown bar at 3 (or 4) different stations. Shade and label 1 fractional unit of each.

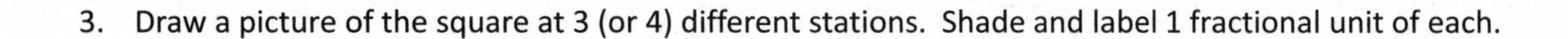

3. Draw a picture of the square at 3 (or 4) different stations. Shade and label 1 fractional unit of each.

4. Draw a picture of the clay at 3 (or 4) different stations. Shade and label 1 fractional unit of each.

5. Draw a picture of the water at 3 (or 4) different stations. Shade and label 1 fractional unit of each.

6. Extension: Draw a picture of the yarn at 3 (or 4) different stations.

Name ______________________________ Date ______________

Each shape is 1 whole. Estimate to equally partition the shape and shade to show the given fraction.

1. 1 fourth

2. 1 fifth ______________________________

3. The shape represents 1 whole. Write the fraction for the shaded part.

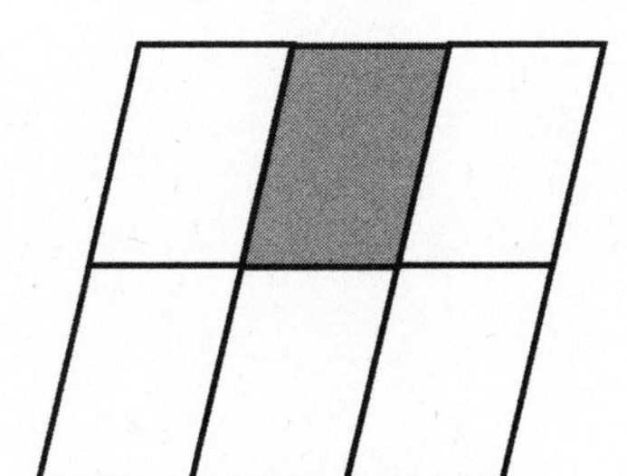

The shaded part is ____________________.

Ms. Browne cut a 6-meter rope into 3 equal-size pieces to make jump ropes. Mr. Ware cut a 5-meter rope into 3 equal size pieces to make jump ropes. Which class has longer jump ropes?

Extension: How long are the jump ropes in Ms. Browne's class?

Read **Draw** **Write**

Name ______________________ Date ______________

1. Fill in the chart. Each image is one whole.

	Total Number of Equal Parts	Total Number of Equal Parts Shaded	Unit Form	Fraction Form
a.	2	1	$\frac{1}{2}$	$\frac{1}{2}$
b.	3	1	$\frac{1}{3}$	$\frac{1}{3}$
c.	4	1	$\frac{1}{4}$	$\frac{1}{4}$
d.	5	1	$\frac{1}{5}$	$\frac{1}{5}$
e.	6	1	$\frac{1}{6}$	$\frac{1}{6}$
f.	8	1	$\frac{1}{8}$	$\frac{1}{8}$

2. Andre's mom baked his 2 favorite cakes for his birthday party. The cakes were the exact same size. Andre cut his first cake into 8 pieces for him and his 7 friends. The picture below shows how he cut it. Did Andre cut the cake into eighths? Explain your answer.

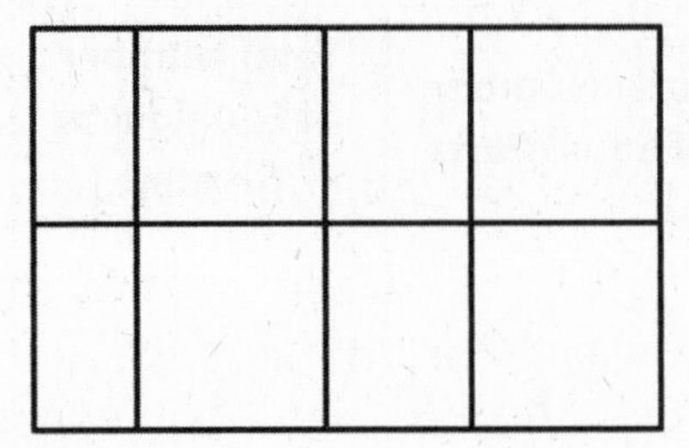

3. Two of Andre's friends came late to his party. They decide they will all share the second cake. Show how Andre can slice the second cake so that he and his nine friends can each get an equal amount with none leftover. What fraction of the second cake will they each receive?

4. Andre thinks it's strange that $\frac{1}{10}$ of the cake would be less than $\frac{1}{8}$ of the cake since ten is bigger than eight. To explain to Andre, draw 2 identical rectangles to represent the cakes. Show 1 tenth shaded on one and 1 eighth shaded on the other. Label the unit fractions and explain to him which slice is bigger.

Name ______________________________ Date ______________

1. Fill in the chart.

	Total Number Equal Parts	Total Number of Equal Parts Shaded	Unit Form	Fraction Form

2. Each image below is 1 whole. Write the fraction that is shaded.

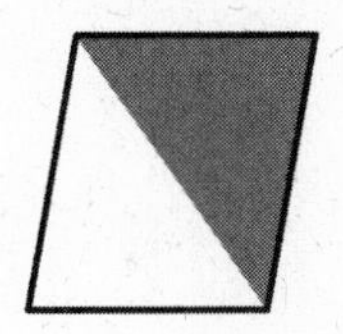

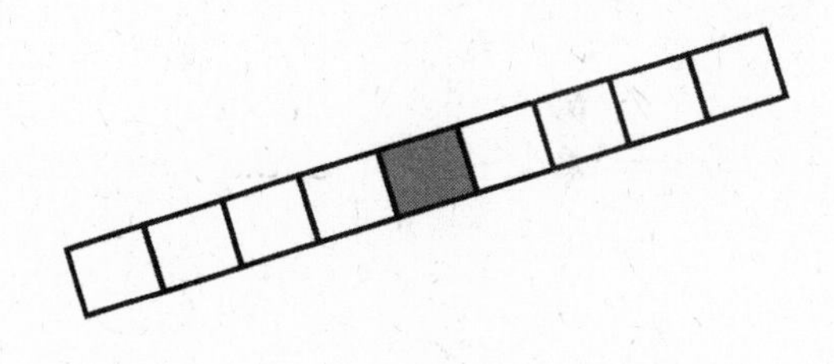

______________ ______________ ______________

3. Draw two identical rectangles. Partition one into 5 equal parts. Partition the other rectangle into 8 equal parts. Label the unit fractions and shade 1 equal part in each rectangle. Use your rectangles to explain why $\frac{1}{5}$ is bigger than $\frac{1}{8}$.

Chloe's dad partitions his garden into 4 equal-sized sections to plant tomatoes, squash, peppers, and cucumbers. What fraction of the garden is available for growing tomatoes?

Extension: Chloe talked her dad into planting beans and lettuce, too. He used equal-sized sections for all the vegetables. What fraction do the tomatoes have now?

Read **Draw** **Write**

Name ______________________________ Date ______________

1. Complete the number sentence. Estimate to partition each strip equally, write the unit fraction inside each unit, and shade the answer.

Sample:

2 thirds = $\frac{2}{3}$

$\frac{1}{3}$	$\frac{1}{3}$	$\frac{1}{3}$

a. 3 fourths =

b. 3 sevenths =

c. 4 fifths =

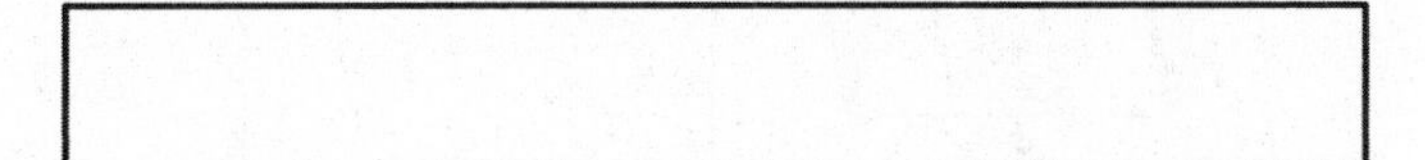

d. 2 sixths =

2. Mr. Stevens bought 8 liters of soda for a party. His guests drank 1 liter.

a. What fraction of the soda did his guests drink?

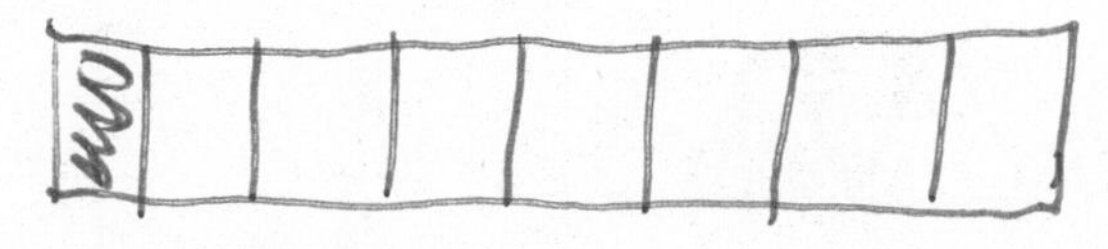

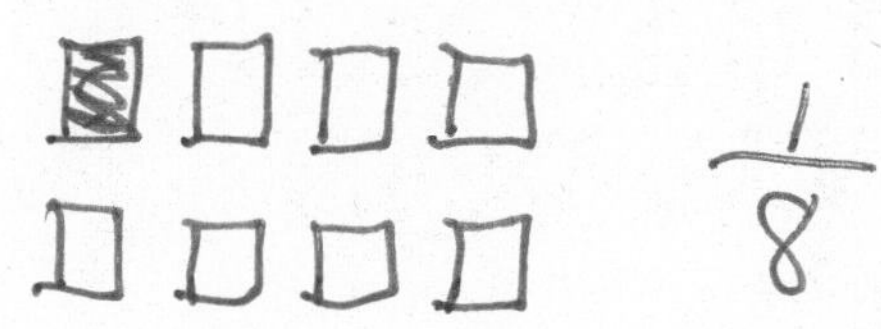

$\frac{1}{8}$

b. What fraction of the soda was left?

3. Fill in the chart.

	Total Number of Equal Parts	Total Number of Shaded Equal Parts	Unit Fraction	Fraction Shaded
Sample:	4	3	$\frac{1}{4}$	$\frac{3}{4}$
a.	9	5	$\frac{1}{9}$	$\frac{5}{9}$
b.	7	3	$\frac{1}{7}$	$\frac{3}{7}$
c.	5	4	$\frac{1}{5}$	$\frac{4}{5}$
d.	6	2	$\frac{1}{6}$	$\frac{2}{6}$
e.	8	8	$\frac{1}{8}$	$\frac{8}{8}=1$

Name ______________________________ Date ______________

1. Complete the number sentence. Estimate to partition the strip equally. Write the unit fraction inside each unit. Shade the answer.

   2 fifths =

2. 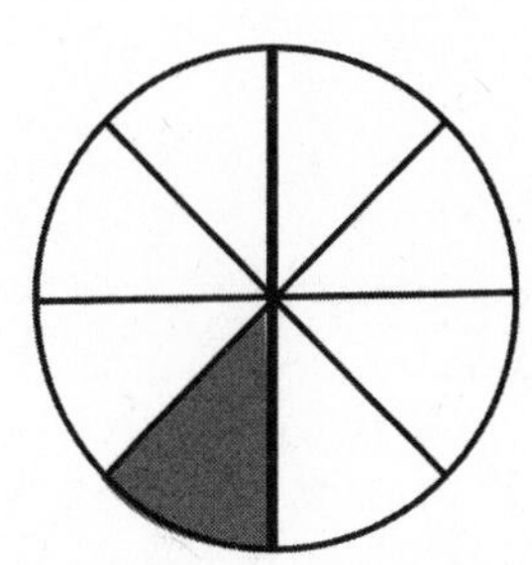

   a. What fraction of the circle is shaded?

   b. What fraction of the circle is not shaded?

3. Complete the chart.

	Total Number of Equal Parts	Total Number of Shaded Equal Parts	Unit Fraction	Fraction Shaded

Robert ate half of the applesauce in a container. He split the remaining applesauce equally into 2 bowls for his mother and sister. Robert said, "I ate 1 half, and each of you gets 1 half."

Is Robert right? Draw a picture to prove your answer.

Extension:

1. What fraction of the applesauce did his mother get?

Read **Draw** **Write**

2. What fraction of the applesauce did Robert's sister eat?

Read **Draw** **Write**

Name ______________________________ Date ______________

Whisper the fraction of the shape that is shaded. Then, match the shape to the amount that is not shaded.

1. 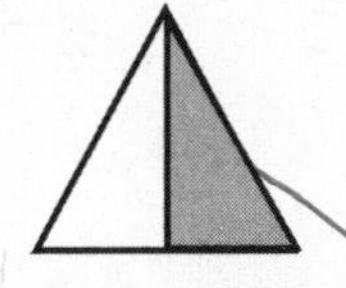

2. 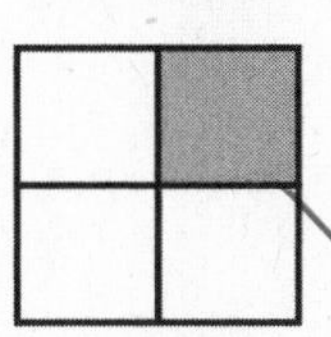

3. 

4. 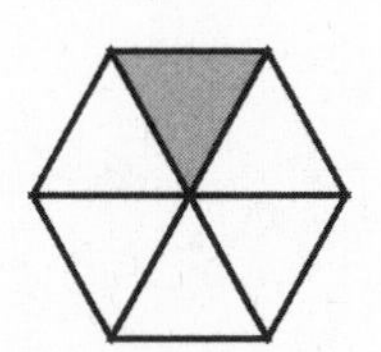

5. 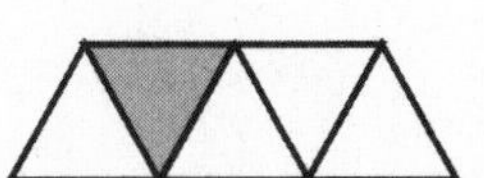

6. 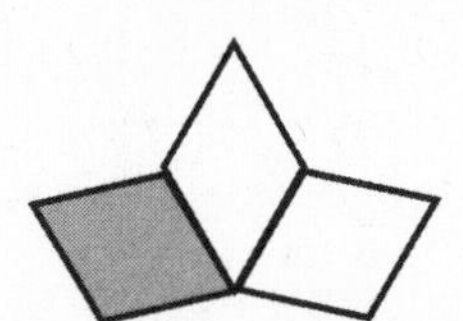

7. 

8. 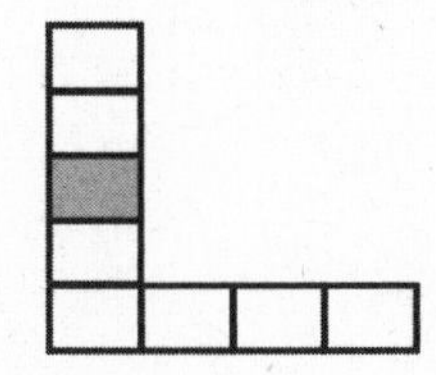

- 2 thirds
- 6 sevenths
- 4 fifths
- 8 ninths
- 1 half
- 5 sixths
- 7 eighths
- 3 fourths

9. a. How many eighths are in 1 whole? 8

   b. How many ninths are in 1 whole? 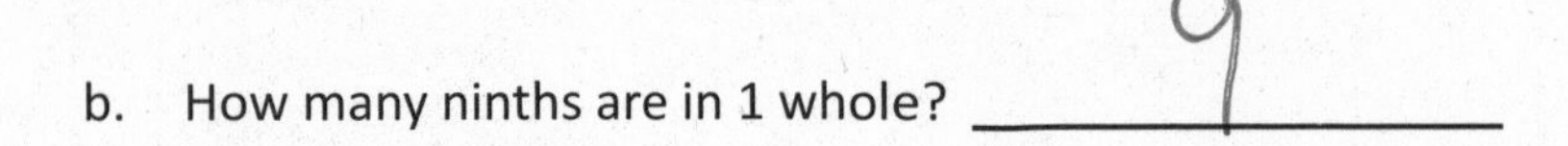

   c. How many twelfths are in 1 whole? 12

10. Each strip represents 1 whole. Write a fraction to label the shaded and unshaded parts.

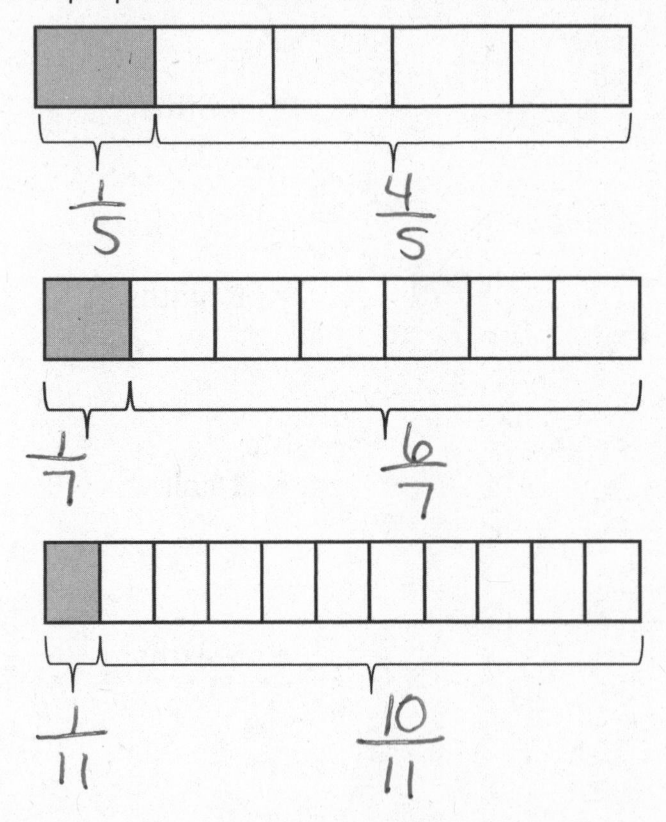

11. Avanti read 1 sixth of her book. What fraction of the book has she not read yet?

Name ____ Date ____________________

1. Write the fraction that is not shaded.

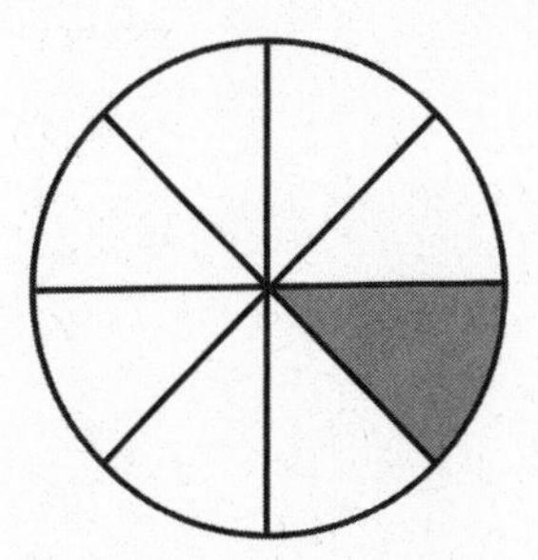

2. There are __________ sixths in 1 whole.

3. The fraction strip is 1 whole. Write fractions to label the shaded and unshaded parts.

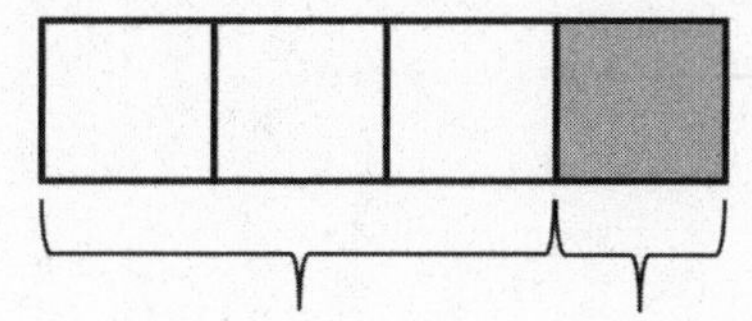

4. Justin mows part of his lawn. Then, his lawnmower runs out of gas. He has not mowed $\frac{9}{1(}$ [illegible] awn. What part of his lawn is mowed?

For breakfast, Mr. Schwartz spent 1 sixth of his money on a coffee and 1 sixth of his money on a bagel. What fraction of his money did Mr. Schwartz spend on breakfast?

Read Draw Write

Name ____________________________ Date ______________

Show a number bond representing what is shaded and unshaded in each of the figures. Draw a different visual model that would be represented by the same number bond.

Sample:

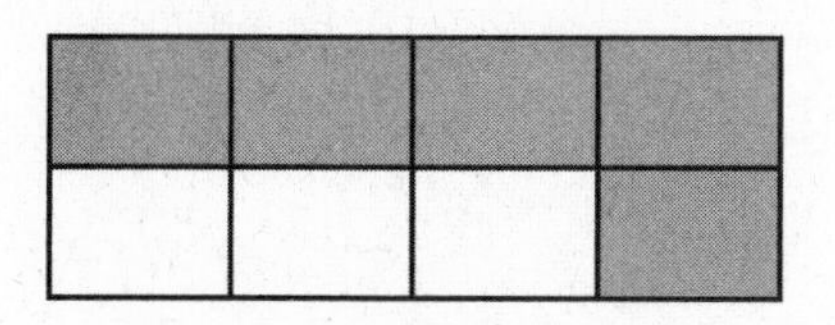

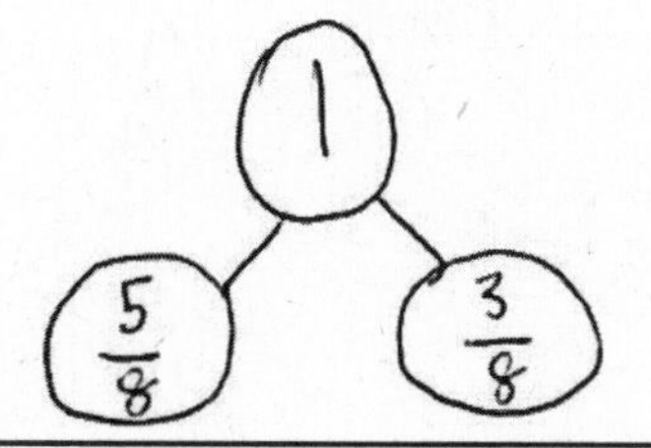

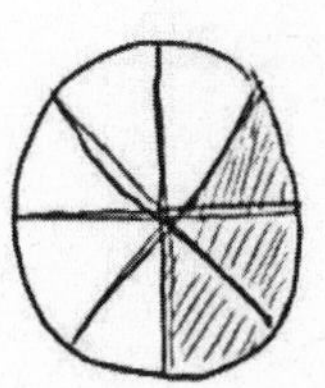

1.

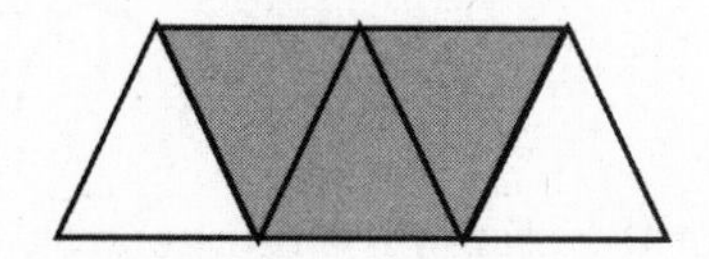

2.

3.

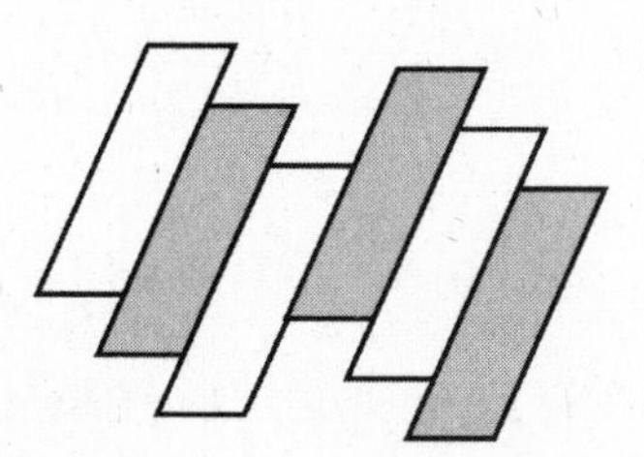

4.

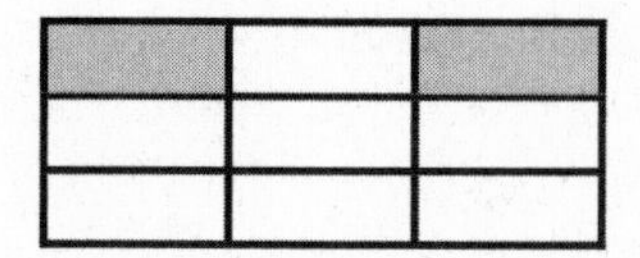

5. Draw a number bond with 2 parts showing the shaded and unshaded fractions of each figure. Decompose both parts of the number bond into unit fractions.

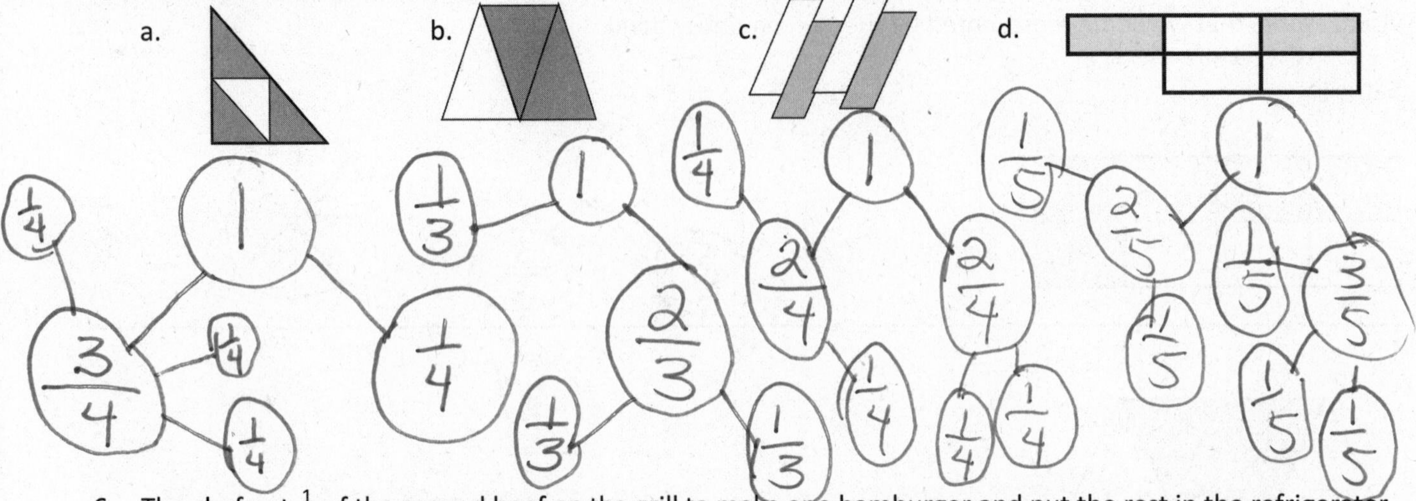

6. The chef put $\frac{1}{4}$ of the ground beef on the grill to make one hamburger and put the rest in the refrigerator. Draw a 2-part number bond showing the fraction of the ground beef on the grill and the fraction in the refrigerator. Draw a visual model of all the ground beef. Shade what is in the refrigerator.

a. What fraction of the ground beef was in the refrigerator?

b. How many more hamburgers can the chef make if he makes them all the same size as the first one?

c. Show the refrigerated ground beef broken into unit fractions on your number bond above.

Name ___________________________________ Date ________________

1. Draw a number bond that shows the shaded and the unshaded parts of the shape below. Then, show each part decomposed into unit fractions.

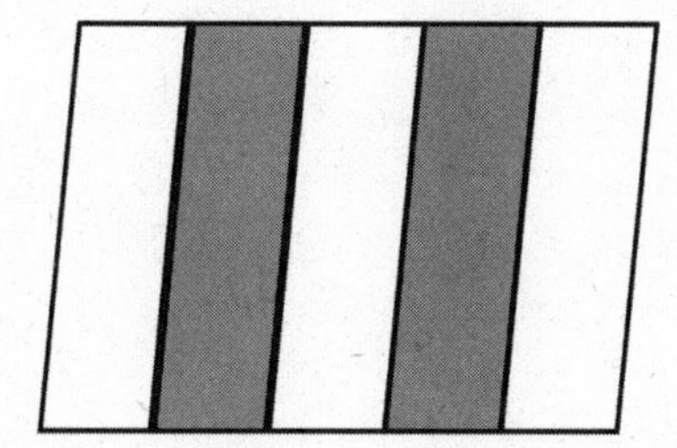

2. Complete the number bond. Draw a shape that has shaded and unshaded parts that match the completed number bond.

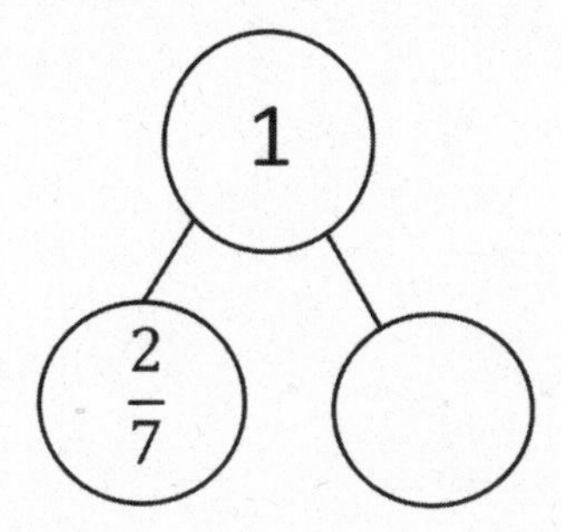

Julianne's friendship bracelet had 8 beads. When it broke, the beads fell off. She could only find 1 bead. To fix her bracelet, what fraction of the beads does she need to buy?

Read **Draw** **Write**

Name ______________________ Date ____________

1. Each figure represents 1 whole. Fill in the chart.

	Unit Fraction	Total Number of Units Shaded	Fraction Shaded
a. Sample:	$\frac{1}{2}$	5	$\frac{5}{2}$
b.	$\frac{1}{8}$	15	$\frac{15}{8}$
c.	$\frac{1}{6}$	14	$\frac{14}{6} = 2\frac{2}{6}$
d.	$\frac{1}{5}$	8	$\frac{8}{5}$
e.	$\frac{1}{4}$	9	$\frac{9}{4}$
f.	$\frac{1}{3}$	7	$\frac{7}{3}$

2. Estimate to draw and shade units on the fraction strips. Solve.

   Sample:

   5 thirds = $\frac{5}{3}$

$\frac{1}{3}$	$\frac{1}{3}$	$\frac{1}{3}$	$\frac{1}{3}$	$\frac{1}{3}$	$\frac{1}{3}$

   a. 8 sixths = $\frac{8}{6}$

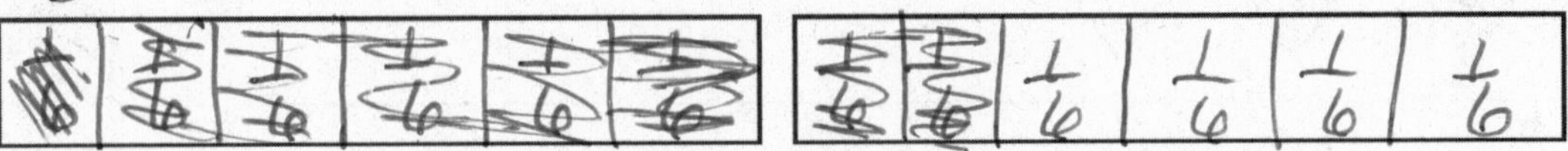

   b. 7 fourths =

   c. ______________________ = $\frac{6}{5}$

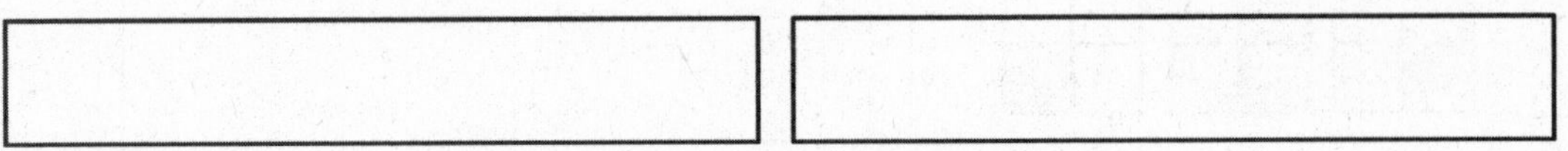

   d. ______________________ = $\frac{5}{2}$

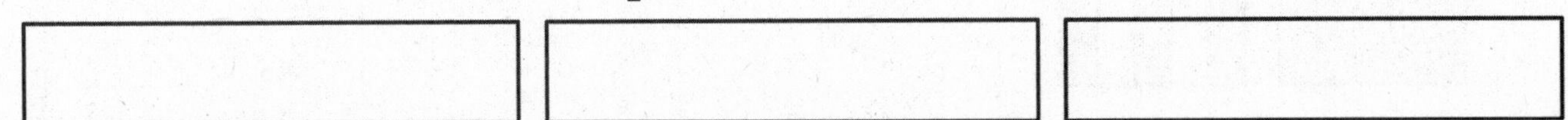

3. Mrs. Jawlik baked 2 pans of brownies. Draw the pans and estimate to partition each pan into 8 equal pieces.

   a. Mrs. Jawlik's children gobbled up 10 pieces. Shade the amount that was eaten.

   b. Write a fraction to show how many pans of brownies her children ate.

Name ______________________________ Date ____________

1. Each shape represents 1 whole. Fill in the chart.

	Unit Fraction	**Total Number of Units Shaded**	**Fraction Shaded**

2. Estimate to draw and shade units on the fraction strips. Solve.

a. 4 thirds =

b. ______________________ $= \frac{10}{4}$

Sarah makes soup. She divides each batch equally into thirds to give away. Each family that she makes soup for gets 1 third of a batch. Sarah needs to make enough soup for 5 families. How much soup does Sarah give away? Write your answer in terms of batches.

Extension: What fraction will be left over for Sarah?

Read **Draw** **Write**

Name ___________________________ Date ___________

1. Each fraction strip is 1 whole. All the fraction strips are equal in length. Color 1 fractional unit in each strip. Then, answer the questions below.

$\frac{1}{2}$

$\frac{1}{4}$

$\frac{1}{8}$

$\frac{1}{3}$

$\frac{1}{6}$

2. Circle *less than* or *greater than.* Whisper the complete sentence.

a. $\frac{1}{2}$ is less than / greater than $\frac{1}{4}$

b. $\frac{1}{6}$ is less than / greater than $\frac{1}{2}$

c. $\frac{1}{3}$ is less than / greater than $\frac{1}{2}$

d. $\frac{1}{3}$ is less than / greater than $\frac{1}{6}$

e. $\frac{1}{8}$ is less than / greater than $\frac{1}{6}$

f. $\frac{1}{8}$ is less than / greater than $\frac{1}{4}$

g. $\frac{1}{2}$ is less than / greater than $\frac{1}{8}$

h. 9 eighths is less than / greater than 2 halves

3. Lily needs $\frac{1}{3}$ cup of oil and $\frac{1}{4}$ cup of water to make muffins. Will Lily use more oil or more water? Explain your answer using pictures, numbers, and words.

4. Use >, <, or = to compare.

a. 1 third ◯ 1 fifth

b. 1 seventh ◯ 1 fourth

c. 1 sixth ◯ $\frac{1}{6}$

d. 1 tenth ◯ $\frac{1}{12}$

e. $\frac{1}{16}$ ◯ 1 eleventh

f. 1 whole ◯ 2 halves

Extension:

g. $\frac{1}{8}$ ◯ 1 eighth ◯ $\frac{1}{6}$ ◯ $\frac{1}{3}$ ◯ 2 halves ◯ 1 whole

5. Your friend Eric says that $\frac{1}{6}$ is greater than $\frac{1}{5}$ because 6 is greater than 5. Is Eric correct? Use words and pictures to explain what happens to the size of a unit fraction when the number of parts gets larger.

Name ______________________________ Date ______________

1. Each fraction strip is 1 whole. All the fraction strips are equal in length. Color 1 fractional unit in each strip. Then, circle the largest fraction and draw a star to the right of the smallest fraction.

$\frac{1}{4}$

$\frac{1}{3}$

$\frac{1}{2}$

2. Use >, <, or = to compare.

a. 1 eighth  1 tenth

b. 1 whole  5 fifths

c. $\frac{1}{7}$  $\frac{1}{6}$

Rachel, Silvia, and Lola each received the same homework assignment and only completed part of it. Rachel completed $\frac{1}{6}$ of her homework, Silvia completed $\frac{1}{2}$ of her homework, and Lola completed $\frac{1}{4}$ of her homework. Write the amount of homework each girl completed from least to greatest. Draw a picture to prove your answer.

$\frac{1}{6}, \frac{1}{4}, \frac{1}{2}$

Read **Draw** **Write**

Name ______________________________ Date ______________

Label the unit fraction. In each blank, draw and label the same whole with a shaded unit fraction that makes the sentence true. There is more than 1 correct way to make the sentence true.

Sample: $\frac{1}{4}$ 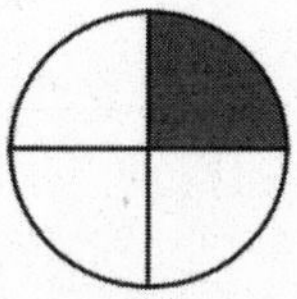	is less than	$\frac{1}{2}$ 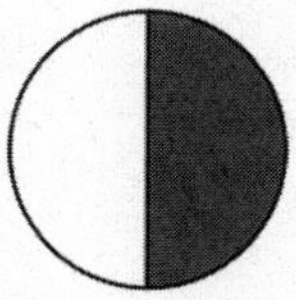
1. 	is greater than	
2. 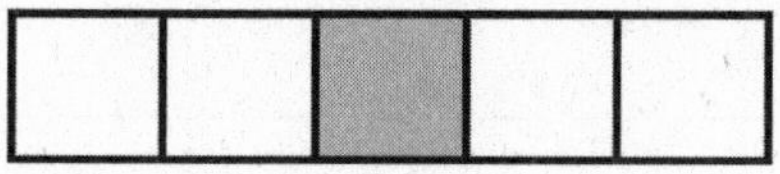	is less than	
3. 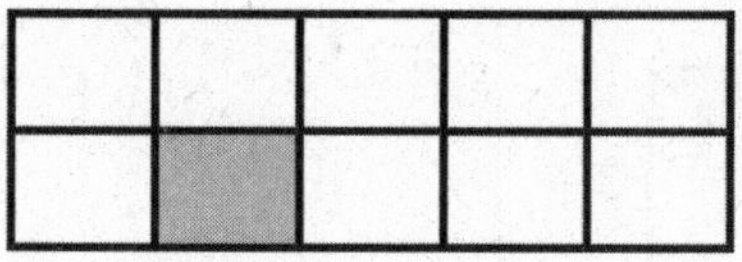	is greater than	
4. 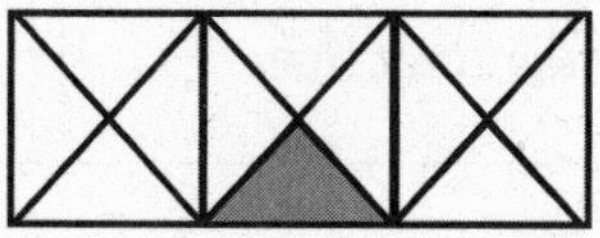	is less than	

5.	is greater than	
6.	is less than	
7.	is greater than	

8. Fill in the blank with a fraction to make the statement true, and draw a matching model.

$\frac{1}{4}$	is less than ☐	$\frac{1}{2}$	is greater than ☐

9. Robert ate $\frac{1}{2}$ of a small pizza. Elizabeth ate $\frac{1}{4}$ of a large pizza. Elizabeth says, "My piece was larger than yours, so that means $\frac{1}{4} > \frac{1}{2}$." Is Elizabeth correct? Explain your answer.

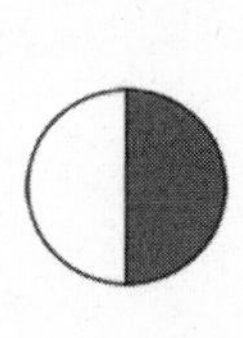

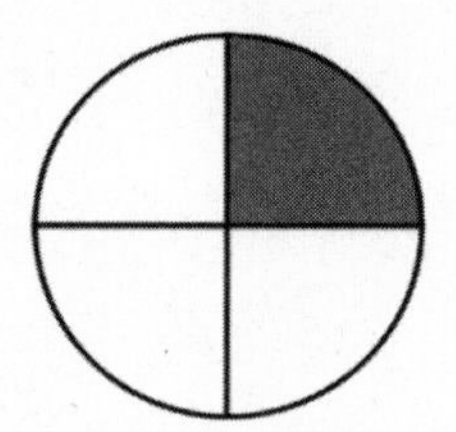

10. Manny and Daniel each ate $\frac{1}{2}$ of his candy, as shown below. Manny said he ate more candy than Daniel because his half is longer. Is he right? Explain your answer.

Manny's Candy Bar

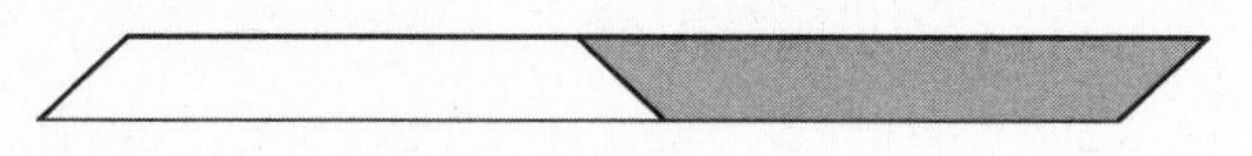

Daniel's Candy Bar

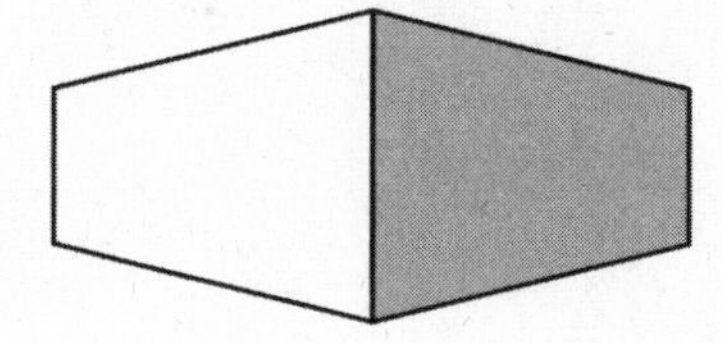

Name ______________________________ Date ______________

1. Fill in the blank with a fraction to make the statement true. Draw a matching model.

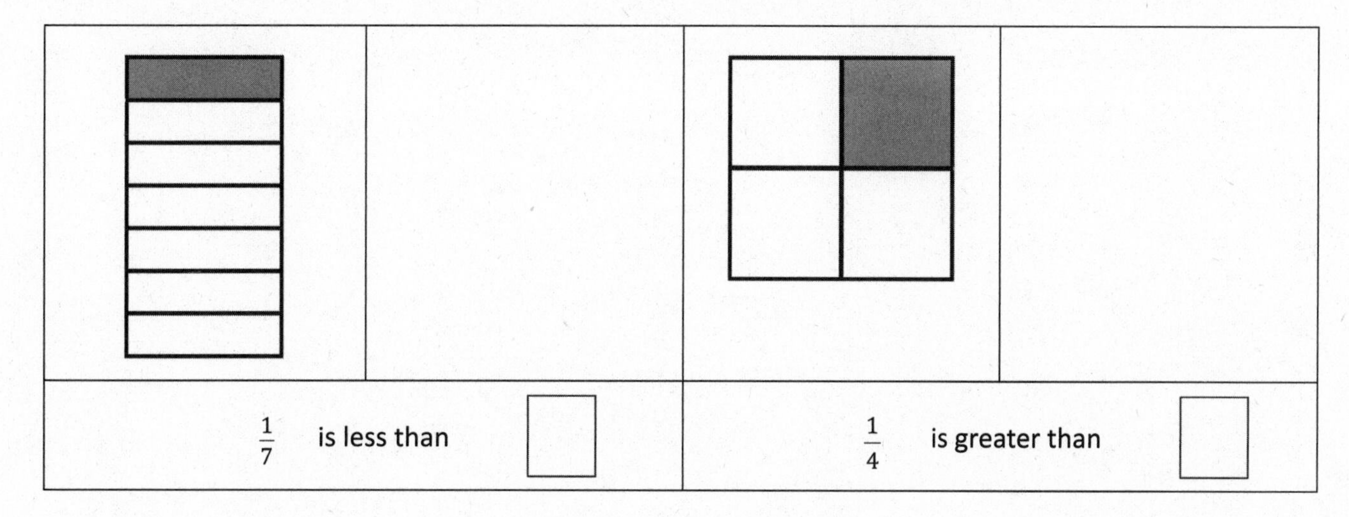

2. Tatiana ate $\frac{1}{2}$ of a small carrot. Louis ate $\frac{1}{4}$ of a large carrot. Who ate more? Use words and pictures to explain your answer.

Jennifer hid half of her birthday money in the dresser drawer. The other half she put in her jewelry box. If she hid $8 in the drawer, how much money did she get for her birthday?

Read Draw Write

Name ______________________________ Date ______________

For each of the following:

- Draw a picture of the designated unit fraction copied to make at least two different wholes.
- Label the unit fractions.
- Label the whole as 1.
- Draw at least one number bond that matches a drawing.

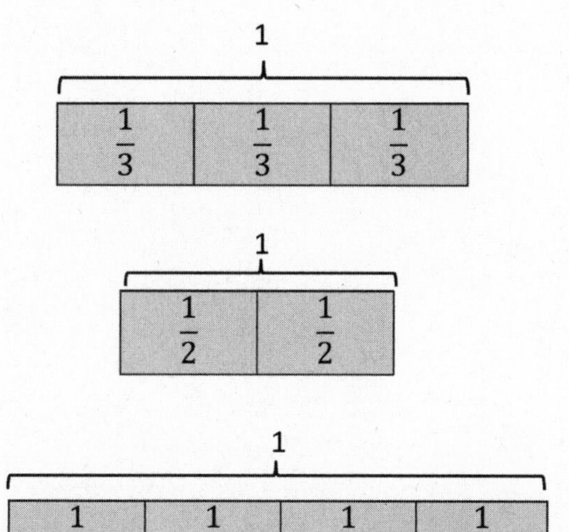

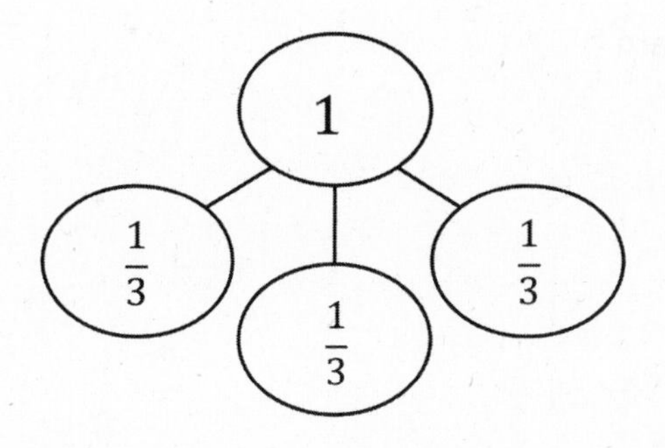

1. Yellow strip

2. Brown strip

3. Orange square

4. Yarn

5. Water

6. Clay

Name ______________________________ Date ______________

Each shape represents the unit fraction. Draw a picture representing a possible whole.

1. 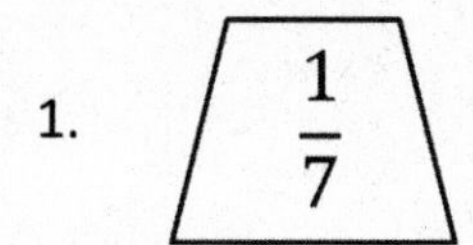

2. 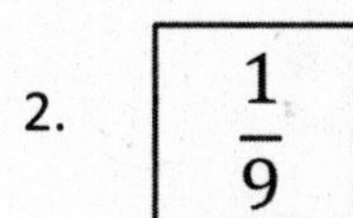

3. Aileen and Jack used the same triangle representing the unit fraction $\frac{1}{4}$ to create 1 whole. Who did it correctly? Explain your answer.

Aileen's Drawing

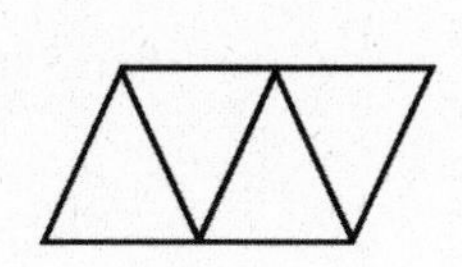

Jack's Drawing

Davis wants to make a picture using 9 square tiles. What fraction of the picture does 1 tile represent? Draw 3 different ways Davis could make his picture.

____

____

____

____

Read **Draw** **Write**

Name ______________________________ Date ______________

The shape represents 1 whole. Write a unit fraction to describe the shaded part.	The shaded part represents 1 whole. Divide 1 whole to show the same unit fraction you wrote in Part (a).
1. a. 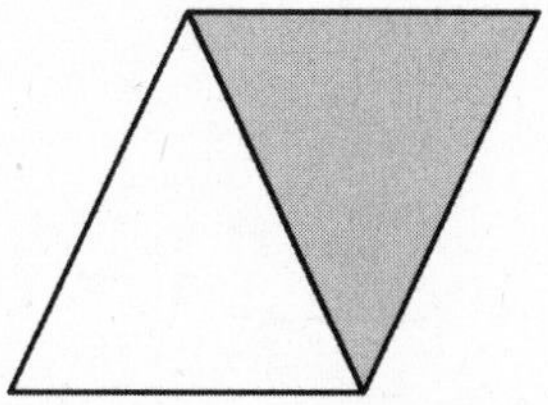	b. 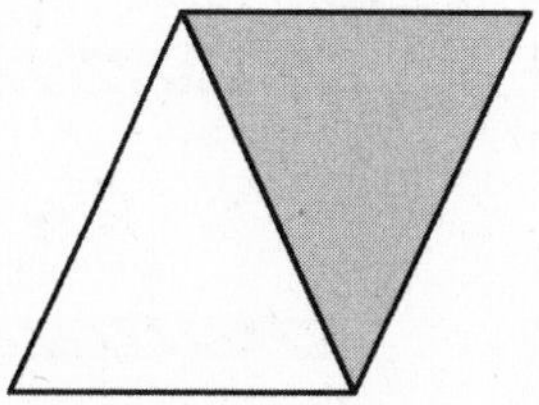
2. a. 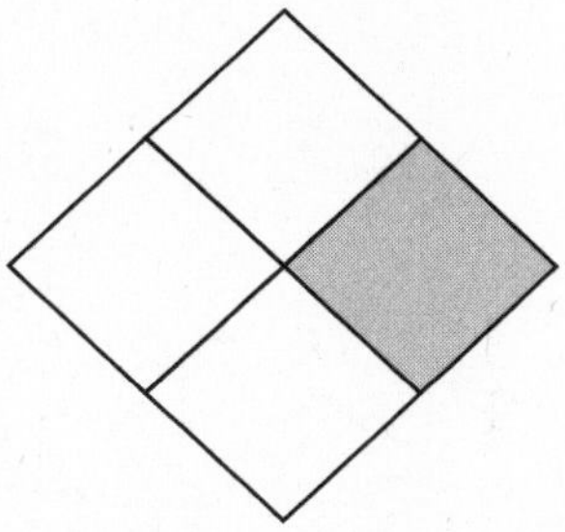	b. 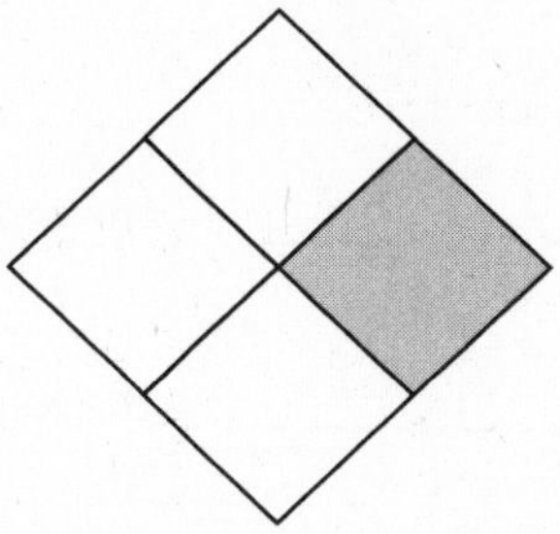
3. a. 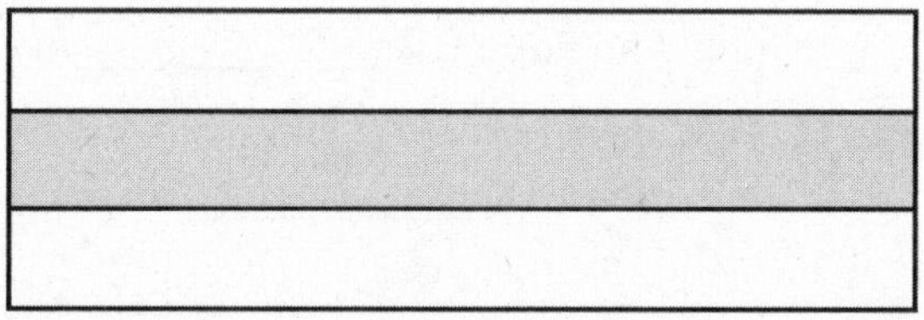	b. 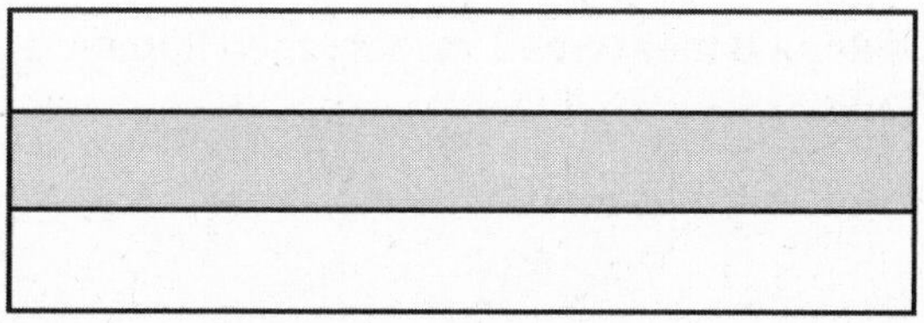
4. a.	b.
5. a. 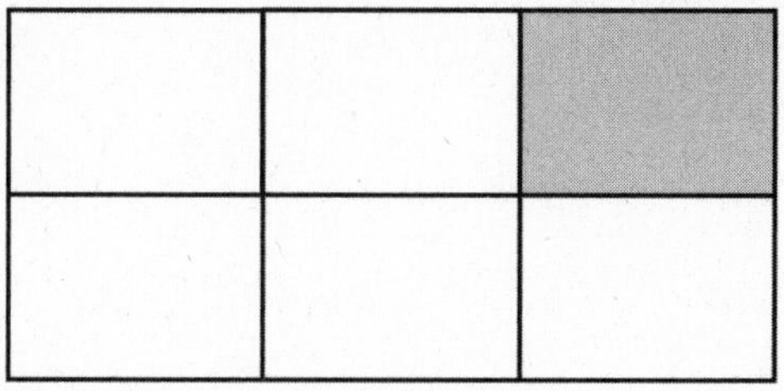	b. 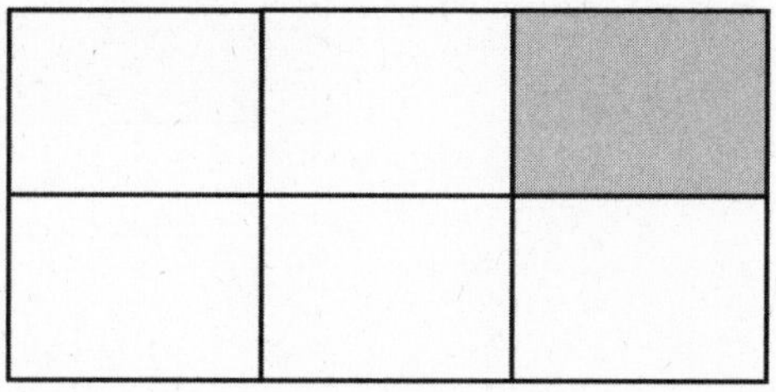

6. Use the diagram below to complete the following statements.

Rope A

Rope B

Rope C

a. Rope ____________ is $\frac{1}{2}$ the length of Rope B.

b. Rope ____________ is $\frac{1}{2}$ the length of Rope A.

c. Rope C is $\frac{1}{4}$ the length of Rope ____________.

d. If Rope B measures 1 m long, then Rope A is ____________ m long, and Rope C is ____________ m long.

e. If Rope A measures 1 m long, Rope B is ____________ m long, and Rope C is ____________ m long.

7. Ms. Fan drew the figure below on the board. She asked the class to name the shaded fraction. Charlie answered $\frac{3}{4}$. Janice answered $\frac{3}{2}$. Jenna thinks they're both right. With whom do you agree? Explain your thinking.

Name ______________________________ Date ______________

Ms. Silverstein asked the class to draw a model showing $\frac{2}{3}$ shaded. Karol and Deb drew the models below. Whose model is correct? Explain how you know.

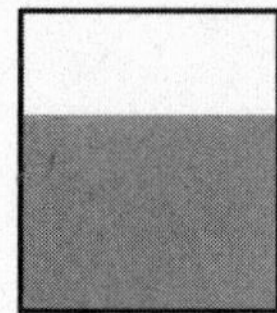

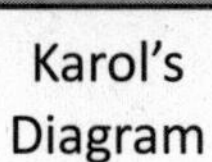

Karol's Diagram

Deb's Diagram

Mr. Ray is knitting a scarf. He says that he has completed 1 fifth of the total length of the scarf. Draw a picture of the final scarf. Label what he has finished and what he still has to make. Draw a number bond with 2 parts to show the fraction he has made and the fraction he has not made.

Read **Draw** **Write**

Name ______________________________ Date ______________

1. Draw a number bond for each fractional unit. Partition the fraction strip to show the unit fractions of the number bond. Use the fraction strip to help you label the fractions on the number line. Be sure to label the fractions at 0 and 1.

a. Halves

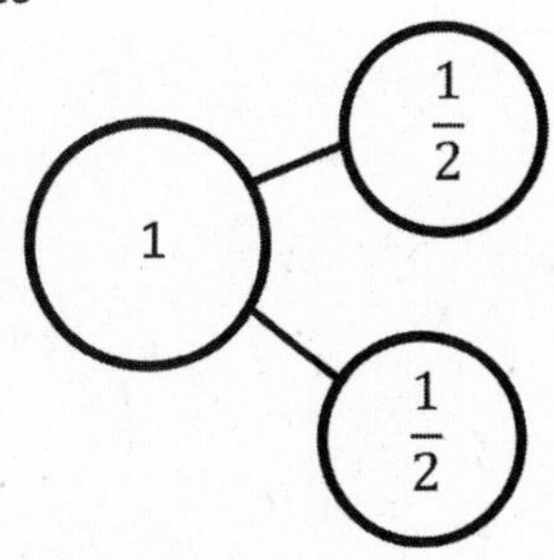

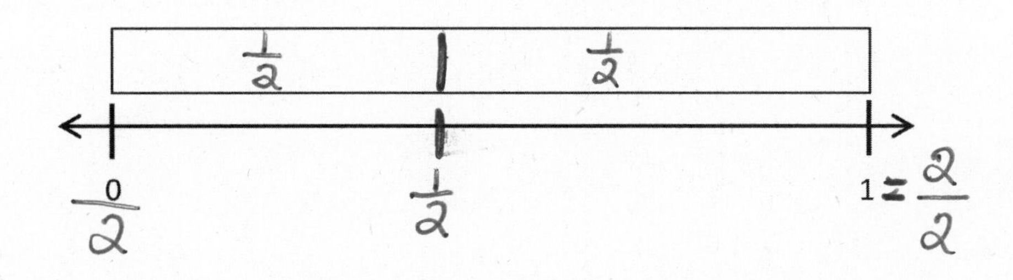

b. Thirds

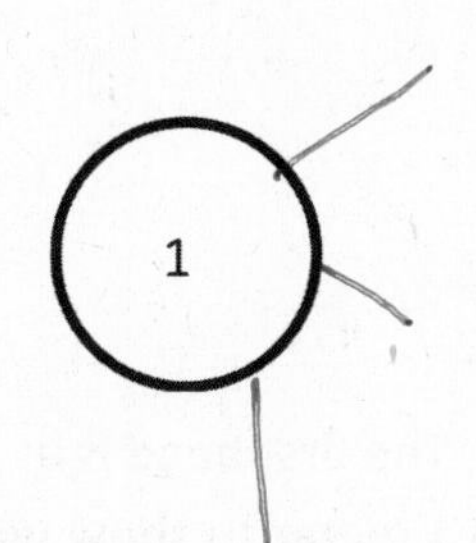

c. Fourths

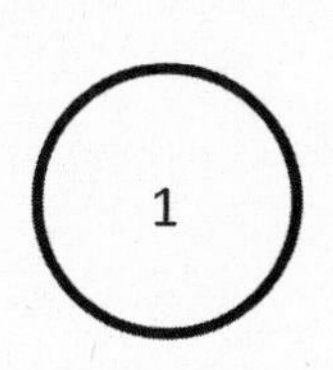

d. Fifths

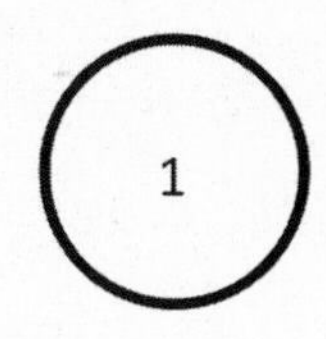

2. Trevor needs to let his puppy outside every quarter (1 fourth) hour to potty train him. Draw and label a number line from 0 hours to 1 hour to show every 1 fourth hour. Include 0 fourths and 4 fourths hour. Label 0 hours and 1 hour, too.

3. A ribbon is 1 meter long. Mrs. Lee wants to sew a bead every $\frac{1}{5}$ meter. The first bead is at $\frac{1}{5}$ meter. The last bead is at 1 meter. Draw and label a number line from 0 meters to 1 meter to show where Mrs. Lee will sew beads. Label all the fractions, including 0 fifths and 5 fifths. Label 0 meters and 1 meter, too.

Name ______________________________________ Date ________________

1. Draw a number bond for the fractional unit. Partition the fraction strip, and draw and label the fractions on the number line. Be sure to label the fractions at 0 and 1.

Sixths (1)

0 1

2. Ms. Metcalf wants to share $1 equally among 5 students. Draw a number bond and a number line to help explain your answer.

   a. What fraction of a dollar will each student get?

   b. How much money will each student get?

In baseball, it is about 30 yards from home plate to first base. The batter got tagged out about halfway to first base. About how many yards from home plate was he when he got tagged out? Draw a number line to show the point where he was when he got tagged out.

Read **Draw** **Write**

Name ____________________________________ Date ____________________

1. Estimate to label the given fractions on the number line. Be sure to label the fractions at 0 and 1. Write the fractions above the number line. Draw a number bond to match your number line.

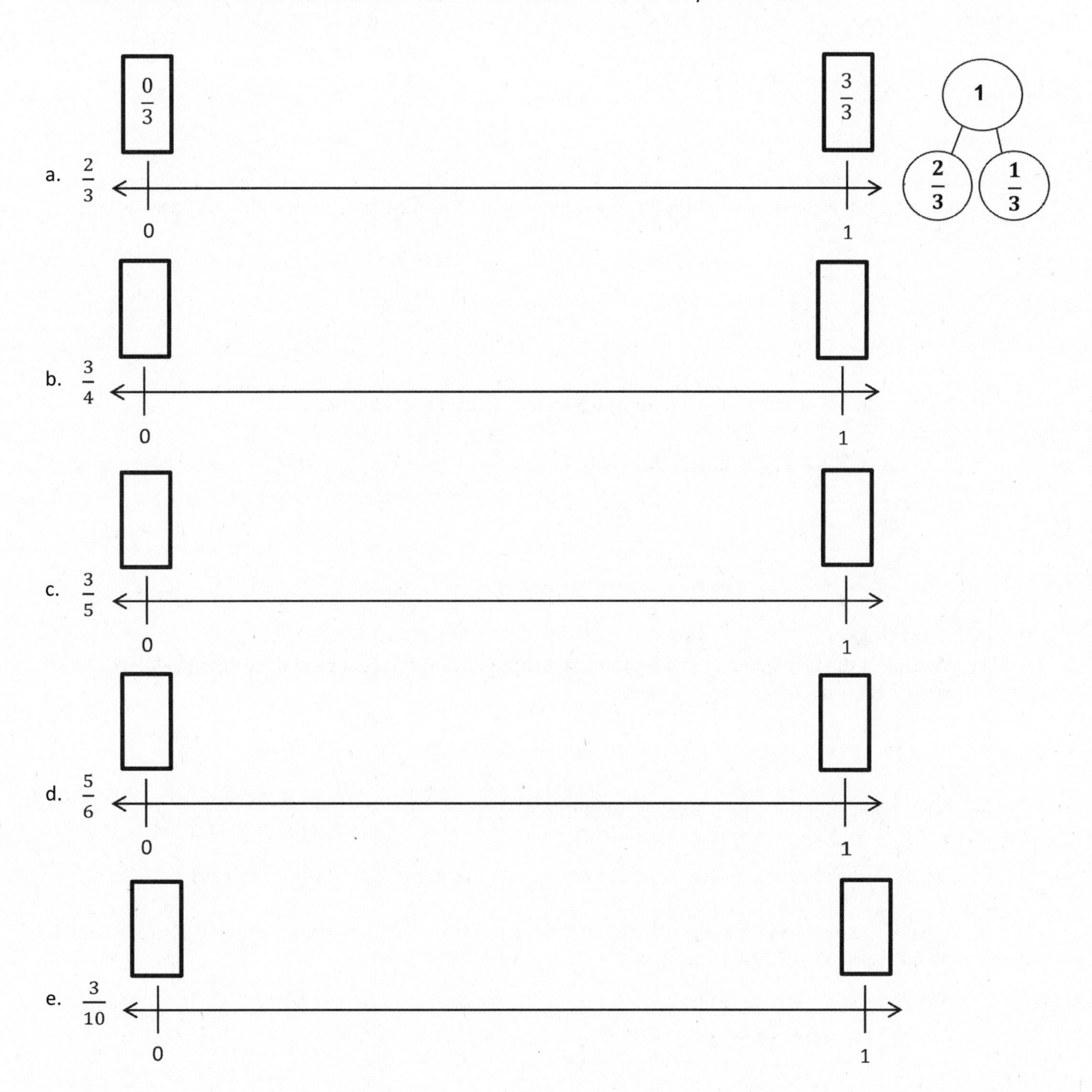

2. Draw a number line. Use a fraction strip to locate 0 and 1. Fold the strip to make 8 equal parts. Use the strip to measure and label your number line with eighths.

   Count up from 0 eighths to 8 eighths on your number line. Touch each number with your finger as you count.

3. For his boat, James stretched out a rope with 5 equally spaced knots as shown.

   a. Starting at the first knot and ending at the last knot, how many equal parts are formed by the 5 knots? Label each fraction at the knot.

   b. What fraction of the rope is labeled at the third knot?

   c. What if the rope had 6 equally spaced knots along the same length? What fraction of the rope would be measured by the first 2 knots?

Name ______________________________________ Date ____________________

1. Estimate to label the given fraction on the number line. Be sure to label the fractions at 0 and 1. Write the fractions above the number line. Draw a number bond to match your number line.

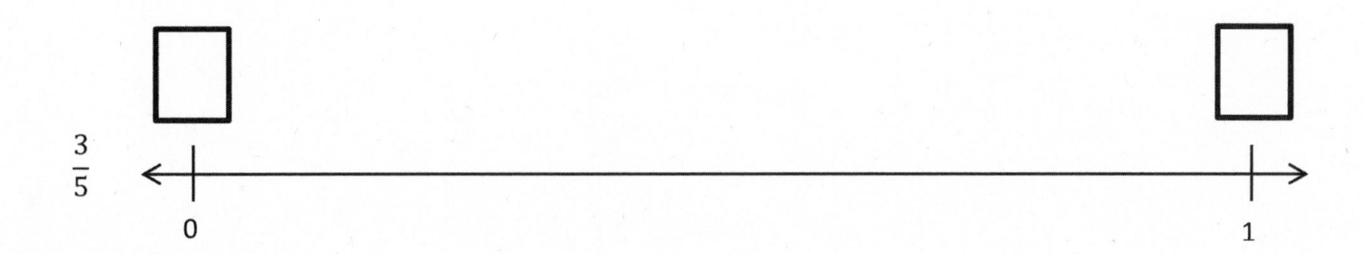

2. Partition the number line. Then, place each fraction on the number line: $\frac{3}{6}$, $\frac{1}{6}$, and $\frac{5}{6}$.

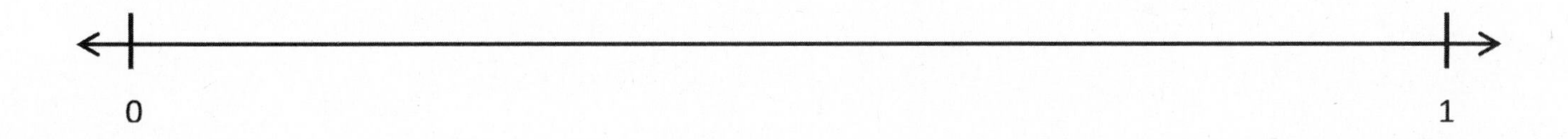

Hannah bought 1 yard of ribbon to wrap 4 small presents. She wants to cut the ribbon into equal parts. Draw and label a number line from 0 yards to 1 yard to show where Hannah will cut the ribbon. Label all the fractions, including 0 fourths and 4 fourths. Also, label 0 yards and 1 yard.

Read **Draw** **Write**

Name ______________________________ Date ______________

1. Estimate to equally partition and label the fractions on the number line. Label the wholes as fractions, and box them. The first one is done for you.

a. halves

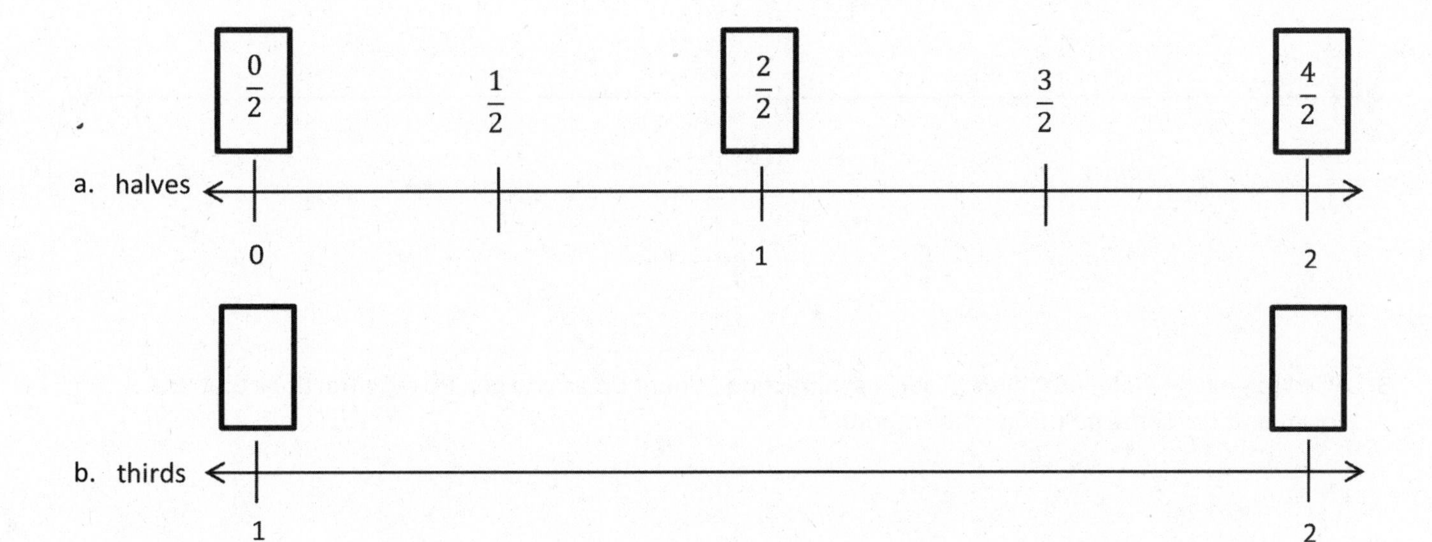

b. thirds

c. halves

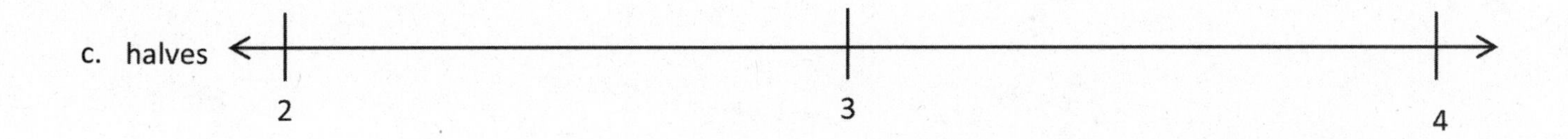

d. fourths

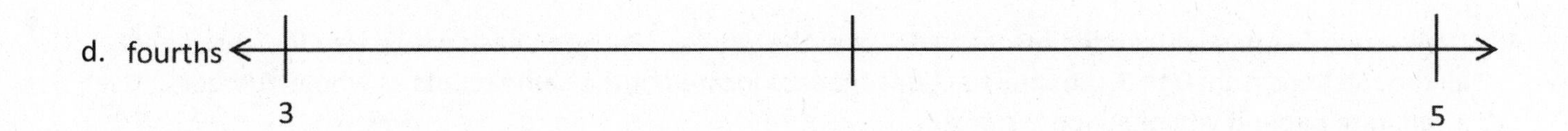

e. thirds

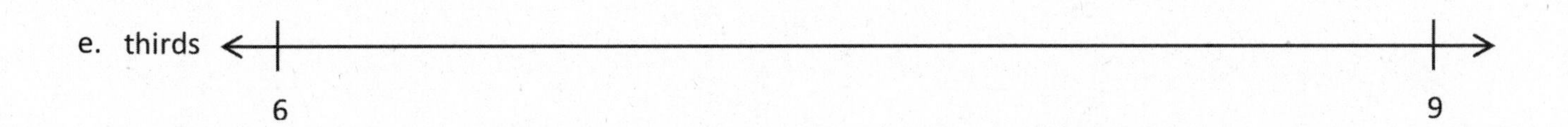

2. Partition each whole into fifths. Label each fraction. Count up as you go. Box the fractions that are located at the same points as whole numbers.

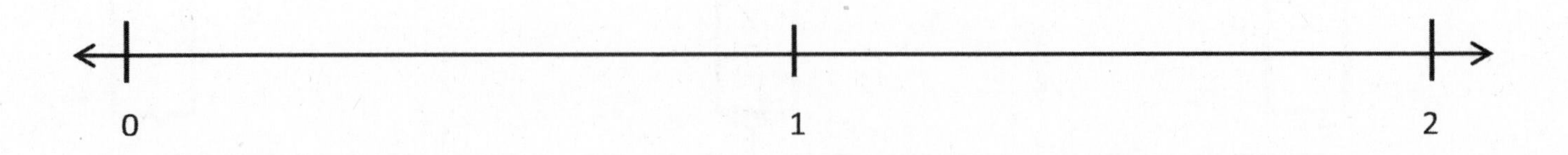

3. Partition each whole into thirds. Label each fraction. Count up as you go. Box the fractions that are located at the same points as whole numbers.

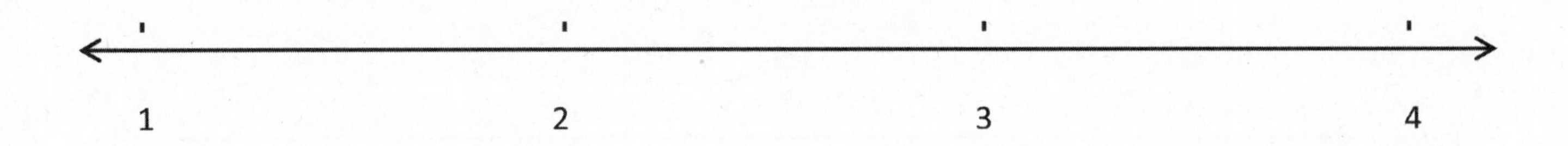

4. Draw a number line with endpoints 0 and 3. Label the wholes. Partition each whole into fourths. Label all the fractions from 0 to 3. Box the fractions that are located at the same points as whole numbers. Use a separate paper if you need more space.

Name ______________________________ Date ______________

1. Estimate to equally partition and label the fractions on the number line. Label the wholes as fractions, and box them.

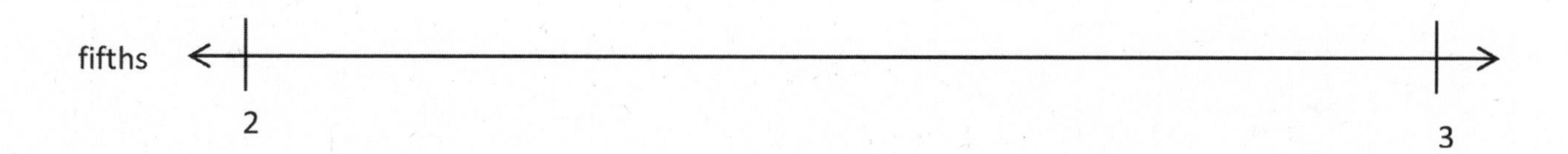

2. Draw a number line with endpoints 0 and 2. Label the wholes. Estimate to partition each whole into sixths, and label them. Box the fractions that are located at the same points as whole numbers.

Sammy sees a black line at the bottom of the pool stretching from one end to the other. She wonders how long it is. The black line is the same length as 9 concrete slabs that make the sidewalk at the edge of the pool. One concrete slab is 5 meters long. What is the length of the black line at the bottom of the pool?

Read **Draw** **Write**

Name ______________________________ Date ______________

1. Locate and label the following fractions on the number line.

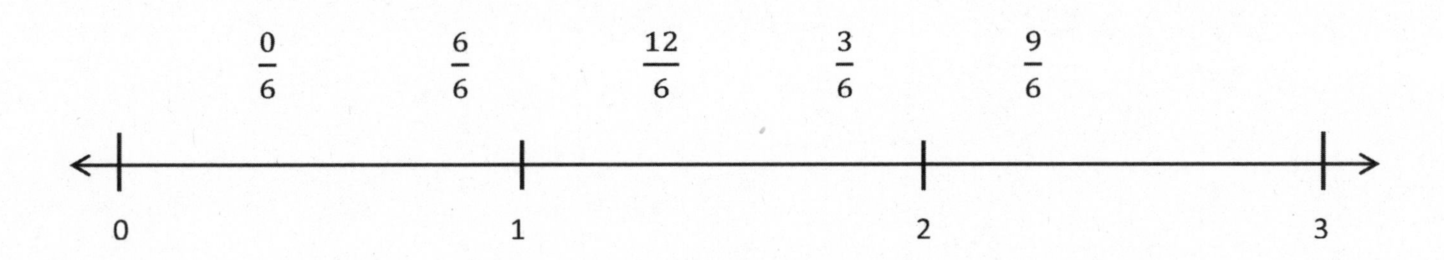

2. Locate and label the following fractions on the number line.

$\frac{8}{4}$ $\frac{6}{4}$ $\frac{12}{4}$ $\frac{16}{4}$ $\frac{4}{4}$

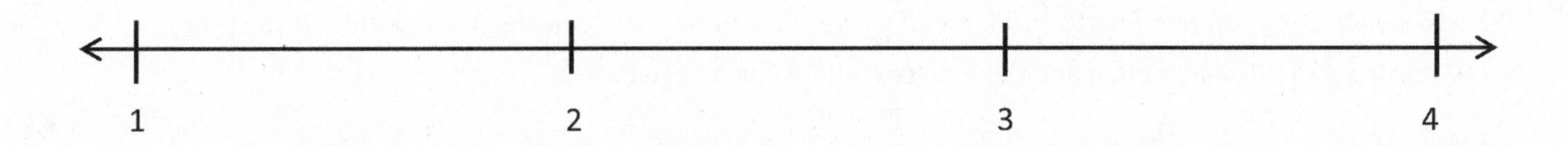

3. Locate and label the following fractions on the number line.

$\frac{18}{3}$ $\frac{14}{3}$ $\frac{9}{3}$ $\frac{11}{3}$ $\frac{6}{3}$

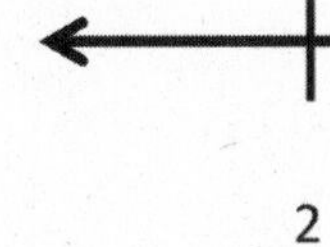

2 3  5 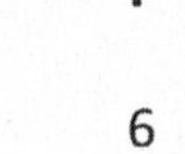

4. For a measurement project in math class, students measured the lengths of their pinky fingers. Alex's measured 2 inches long. Jerimiah's pinky finger was $\frac{7}{4}$ inches long. Whose finger is longer? Draw a number line to help prove your answer.

5. Marcy ran 4 kilometers after school. She stopped to tie her shoelace at $\frac{7}{5}$ kilometers. Then, she stopped to switch songs on her iPod at $\frac{12}{5}$ kilometers. Draw a number line showing Marcy's run. Include her starting and finishing points and the 2 places where she stopped.

Name ______________________________ Date ______________

1. Locate and label the following fractions on the number line.

$\frac{7}{3}$ $\frac{2}{3}$ $\frac{4}{3}$

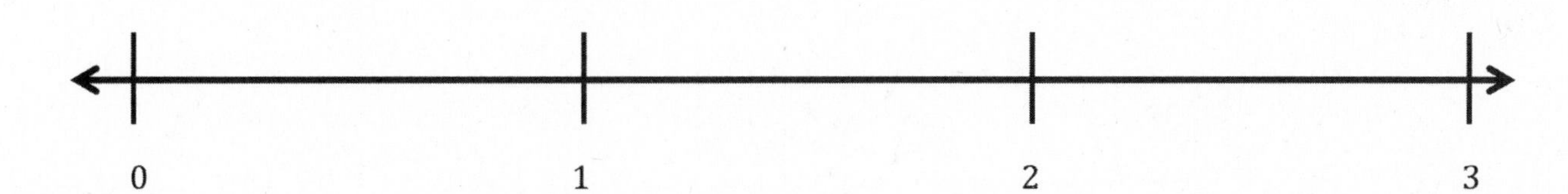

2. Katie bought 2 one-gallon bottles of juice for a party. Her guests drank $\frac{6}{4}$ gallons of juice. What fraction of a gallon of juice is left over? Draw a number line to show, and explain your answer.

Third-grade students are growing peppers. The student with the longest pepper wins the Green Thumb award. Jackson's pepper measured 3 inches long. Drew's measured $\frac{10}{4}$ inches long. Who won the award? Draw a number line to help prove your answer.

Read **Draw** **Write**

Name ____________________ Date ____________

Place the two fractions on the number line. Circle the fraction with the distance closest to 0. Then, compare using >, <, or =. The first problem is done for you.

1. 
$\frac{1}{4}$ < $\frac{3}{4}$

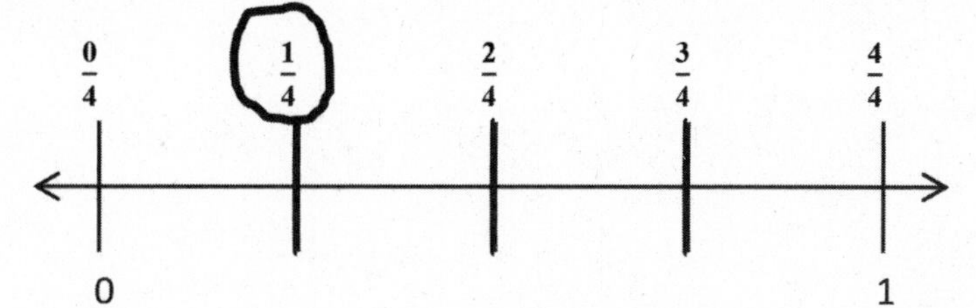

2. 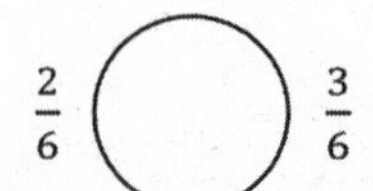
$\frac{2}{6}$ ◯ $\frac{3}{6}$

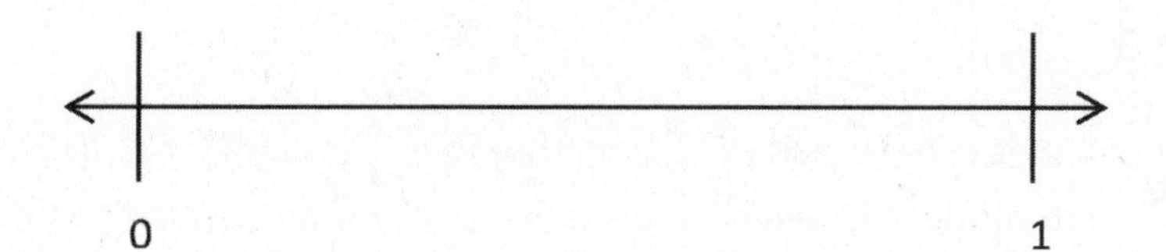

3. 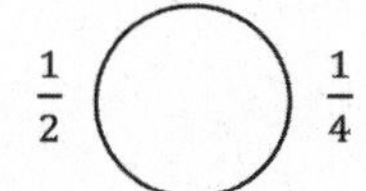
$\frac{1}{2}$ ◯ $\frac{1}{4}$

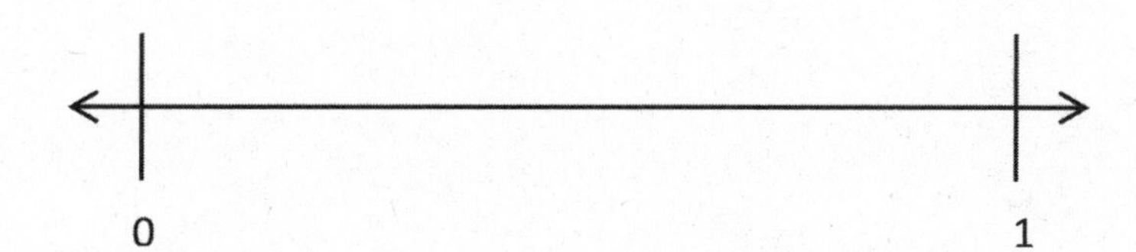

4. 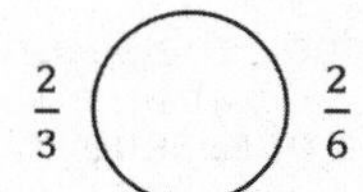
$\frac{2}{3}$ ◯ $\frac{2}{6}$

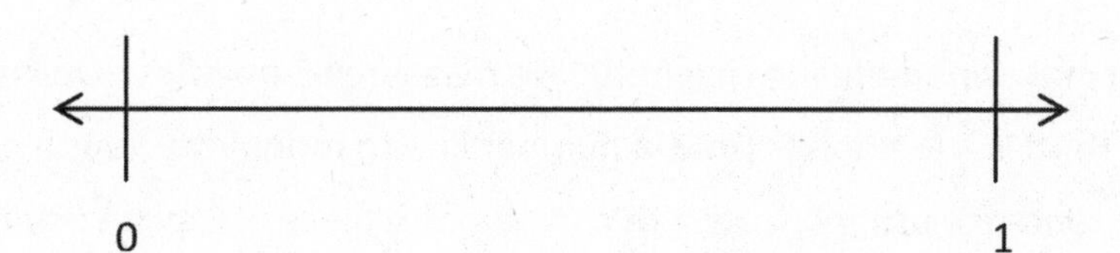

5. 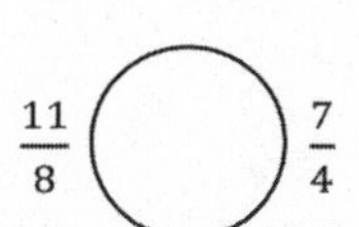
$\frac{11}{8}$ ◯ $\frac{7}{4}$

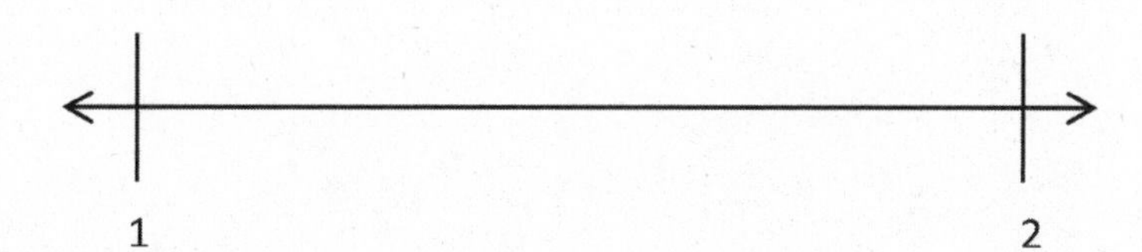

6. JoAnn and Lupe live straight down the street from their school. JoAnn walks $\frac{5}{6}$ miles and Lupe walks $\frac{7}{8}$ miles home from school every day. Draw a number line to model how far each girl walks. Who walks the least? Explain how you know using pictures, numbers, and words.

7. Cheryl cuts 2 pieces of thread. The blue thread is $\frac{5}{4}$ meters long. The red thread is $\frac{4}{5}$ meters long. Draw a number line to model the length of each piece of thread. Which piece of thread is shorter? Explain how you know using pictures, numbers, and words.

8. Brandon makes homemade spaghetti. He measures 3 noodles. One measures $\frac{7}{8}$ feet, the second is $\frac{7}{4}$ feet, and the third is $\frac{4}{2}$ feet long. Draw a number line to model the length of each piece of spaghetti. Write a number sentence using <, >, or = to compare the pieces. Explain using pictures, numbers, and words.

Name ______________________________ Date ______________

Place the two fractions on the number line. Circle the fraction with the distance closest to 0. Then, compare using >, <, or =.

1. $\frac{3}{5}$ ◯ $\frac{1}{5}$

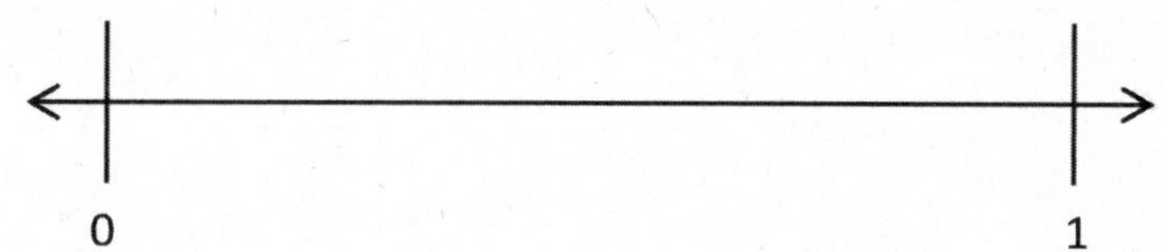

2. $\frac{1}{2}$ ◯ $\frac{3}{4}$

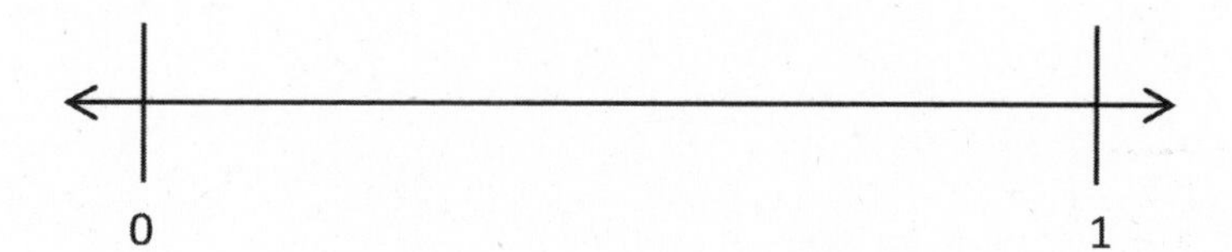

3. Mr. Brady draws a fraction on the board. Ken says it's $\frac{2}{3}$, and Dan said it's $\frac{3}{2}$. Do both of these fractions mean the same thing? If not, which fraction is larger? Draw a number line to model $\frac{2}{3}$ and $\frac{3}{2}$. Use words, pictures, and numbers to explain your comparison.

Thomas has 2 sheets of paper. He wants to punch 4 equally spaced holes along the edge of each sheet. Draw Thomas's 2 sheets of paper next to each other so the ends meet. Label a number line from 0 at the start of his first paper to 2 at the end of his second paper. Show Thomas where to hole-punch his papers and label the fractions. What fraction is labeled at the eighth hole?

Read **Draw** **Write**

Name ______________________________ Date ______________

1. Divide each number line into the given fractional unit. Then, place the fractions. Write each whole as a fraction.

a. halves $\frac{3}{2}$ $\frac{5}{2}$ $\frac{4}{2}$

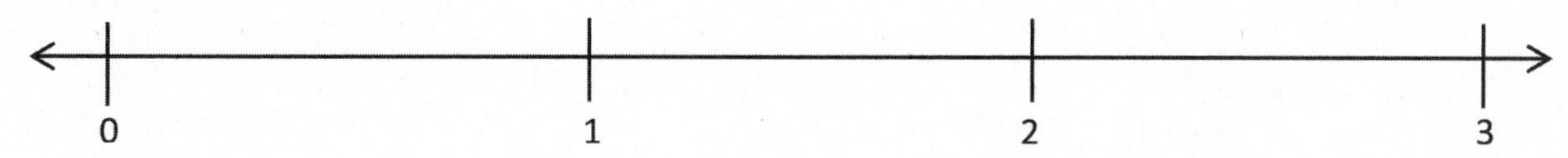

b. fourths $\frac{9}{4}$ $\frac{11}{4}$ $\frac{6}{4}$

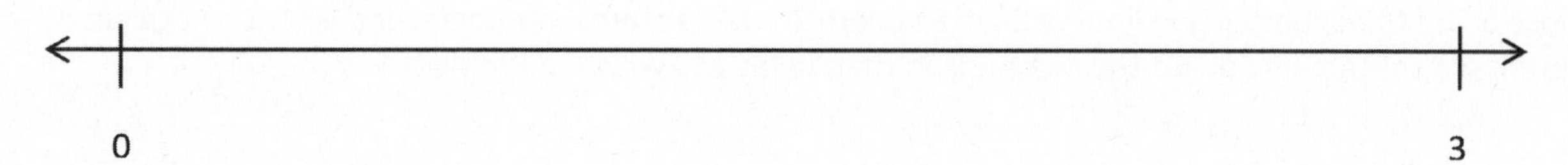

c. eighths $\frac{24}{8}$ $\frac{19}{8}$ $\frac{16}{8}$

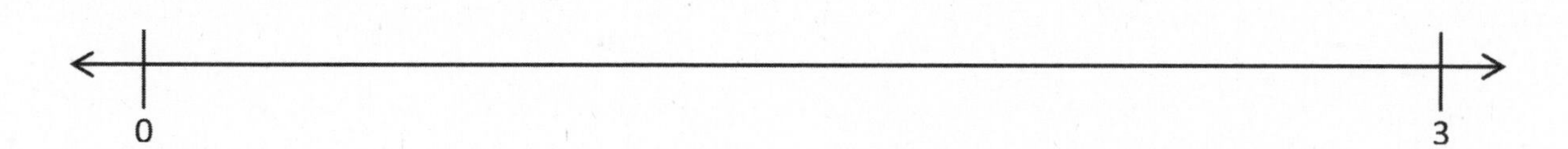

2. Use the number lines above to compare the following fractions using >, <, or =.

$\frac{6}{4}$ ◯ $\frac{9}{4}$ $\frac{3}{2}$ ◯ $\frac{5}{2}$ $\frac{19}{8}$ ◯ $\frac{16}{8}$

$\frac{16}{8}$ ◯ $\frac{3}{2}$ $\frac{9}{4}$ ◯ $\frac{19}{8}$ $\frac{4}{2}$ ◯ $\frac{16}{8}$

$\frac{6}{4}$ ◯ $\frac{16}{8}$ $\frac{5}{2}$ ◯ $\frac{9}{4}$ $\frac{24}{8}$ ◯ $\frac{11}{4}$

3. Choose a *greater than* comparison you made in Problem 2. Use pictures, numbers, and words to explain how you made that comparison.

4. Choose a *less than* comparison you made in Problem 2. Use pictures, numbers, and words to explain a different way of thinking about the comparison than what you wrote in Problem 3.

5. Choose an *equal to* comparison you made in Problem 2. Use pictures, numbers, and words to explain two ways that you can prove your comparison is true.

Name ______________________________ Date ____________

1. Divide the number line into the given fractional unit. Then, place the fractions. Write each whole as a fraction.

   fourths $\frac{2}{4}$ $\frac{10}{4}$ $\frac{7}{4}$

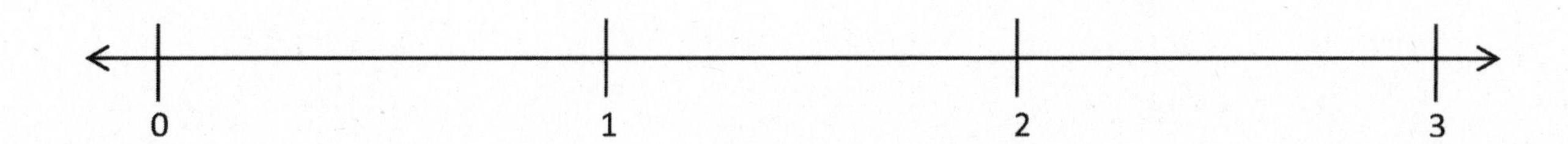

2. Use the number line above to compare the following fractions using >, <, or =.

   $\frac{3}{4}$ ◯ $\frac{5}{4}$ $\frac{7}{4}$ ◯ $\frac{4}{4}$ 3 ◯ $\frac{6}{4}$

3. Use the number line from Problem 1. Which is larger: 2 wholes or $\frac{9}{4}$? Use words, pictures, and numbers to explain your answer.

Max ate $\frac{2}{3}$ of his pizza for lunch. He wanted to eat a small snack in the afternoon, so he cut the leftover pizza in half and ate 1 slice. How much of the pizza was left? Draw a picture to help you think about the pizza.

Read **Draw** **Write**

Name ______________________________ Date ______________

1. Label what fraction of each shape is shaded. Then, circle the fractions that are equal.

a.

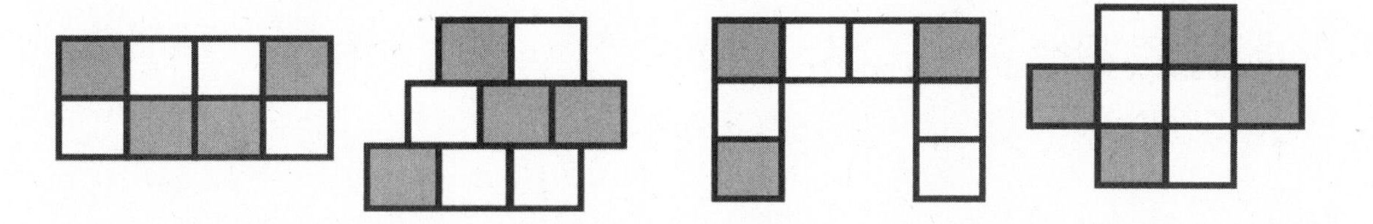

b.

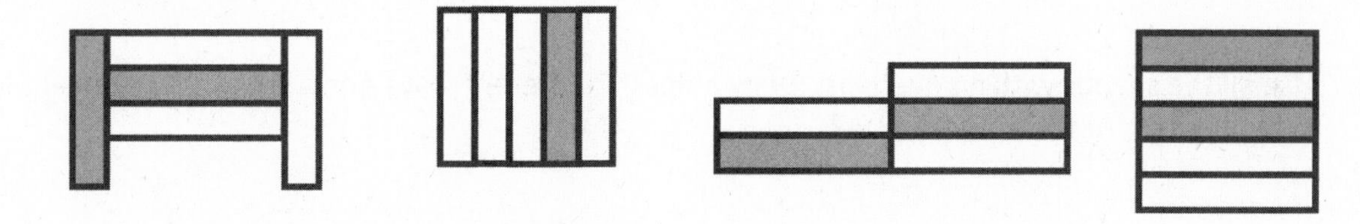

c.

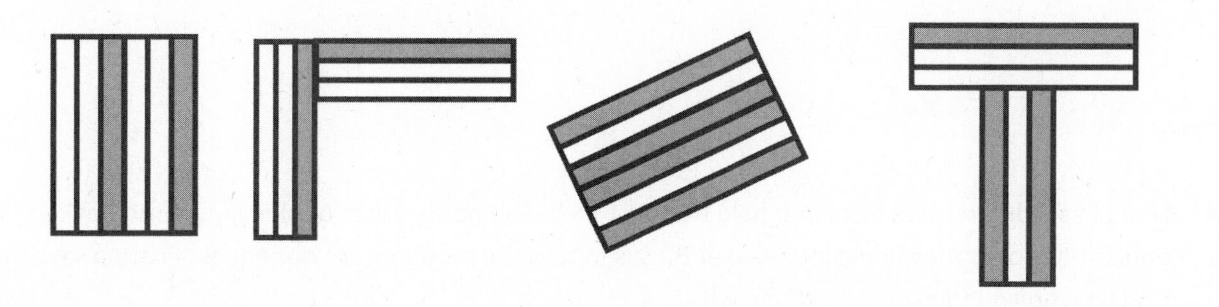

2. Label the shaded fraction. Draw 2 different representations of the same fractional amount.

a.

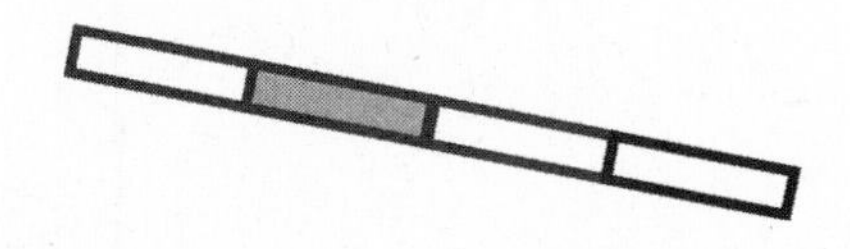

b.

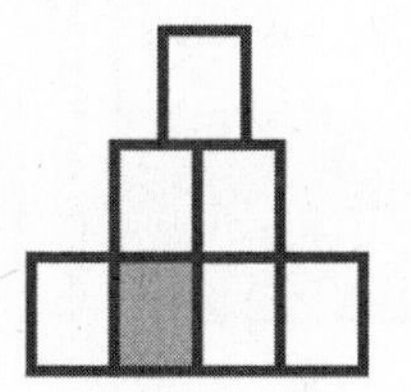

3. Ann has 6 small square pieces of paper. 2 squares are grey. Ann cuts the 2 grey squares in half with a diagonal line from one corner to the other.

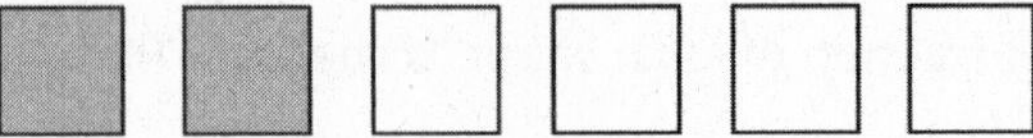

a. What shapes does she have now?

b. How many of each shape does she have?

c. Use all the shapes with no overlaps. Draw at least 2 different ways Ann's set of shapes might look. What fraction of the figure is grey?

4. Laura has 2 different beakers that hold exactly 1 liter. She pours $\frac{1}{2}$ liter of blue liquid into Beaker A. She pours $\frac{1}{2}$ liter of orange liquid into Beaker B. Susan says the amounts are not equal. Cristina says they are. Explain who you think is correct and why.

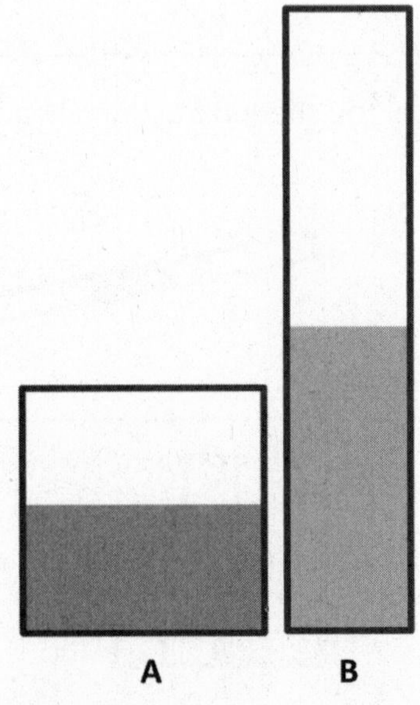

Name ____ Date ____________________

1. Label what fraction of the figure is shaded. Then, circle the fractions that are equal.

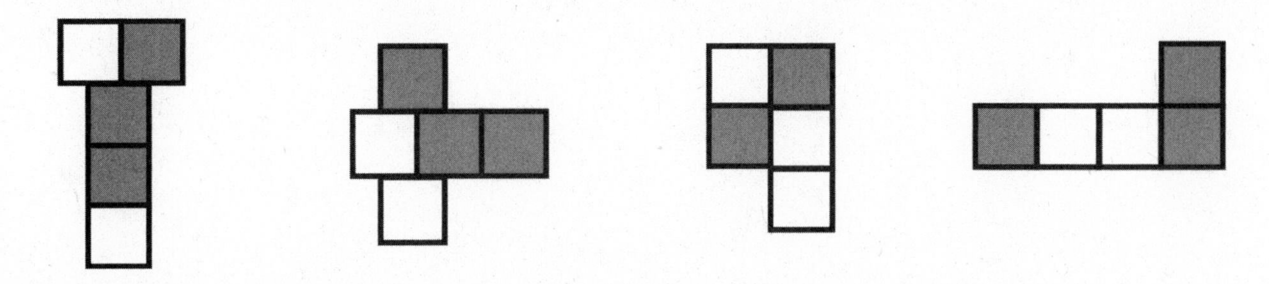

2. Label the shaded fraction. Draw 2 different representations of the same fractional amount.

a.

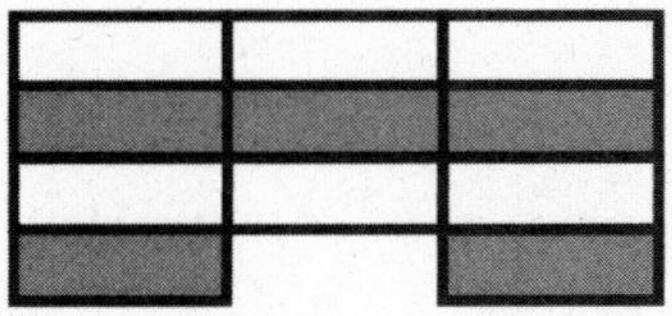

b.

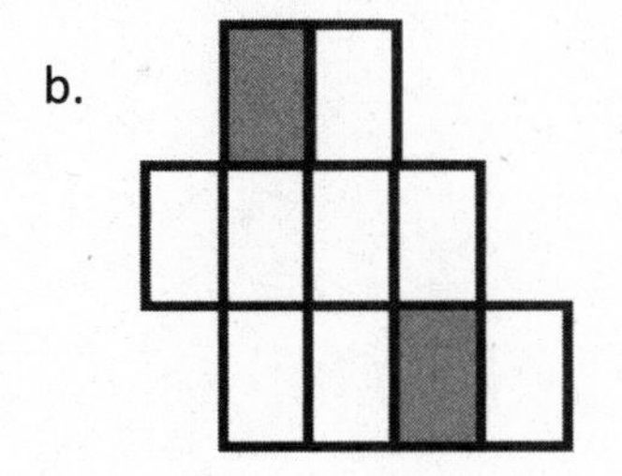

Dorothea is training to run a 2-mile race. She marks off her starting point and the finish line. To track her progress, she places a mark at 1 mile. She then places a mark halfway between her starting position and 1 mile, and another mark halfway between 1 mile and the finish line.

a. Draw and label a number line to show the points Dorothea marks along her run.

b. What fractional unit does Dorothea make as she marks the points on her run?

Read **Draw** **Write**

c. What fraction of her run has she completed when she reaches the third marker?

Read **Draw** **Write**

Name ______________________________ Date ______________

1. Use the fractional units on the left to count up on the number line. Label the missing fractions on the blanks.

$\frac{1}{2}$ $\frac{2}{2}$ $\frac{4}{2}$

halves

0 1 2

fourths $\frac{0}{4}$ ___ ___ $\frac{3}{4}$ ___ $\frac{5}{4}$ ___ $\frac{7}{4}$ ___

2. Use the number lines above to:
   - Color fractions equal to 1 half blue.
   - Color fractions equal to 1 yellow.
   - Color fractions equal to 3 halves green.
   - Color fractions equal to 2 red.

3. Use the number lines above to make the number sentences true.

$$\frac{2}{4} = \frac{\ }{6} \qquad \frac{6}{6} = \frac{2}{\ } = \frac{\ }{\ } \qquad \frac{3}{2} = \frac{\ }{6} = \frac{6}{\ }$$

4. Jack and Jill use rain gauges the same size and shape to measure rain on the top of a hill. Jack uses a rain gauge marked in fourths of an inch. Jill's gauge measures rain in eighths of an inch. On Thursday, Jack's gauge measured $\frac{2}{4}$ inches of rain. They both had the same amount of water, so what was the reading on Jill's gauge Thursday? Draw a number line to help explain your thinking.

5. Jack and Jill's baby brother Rosco also had a gauge the same size and shape on the same hill. He told Jack and Jill that there had been $\frac{1}{2}$ inch of rain on Thursday. Is he right? Why or why not? Use words and a number line to explain your answer.

Name ____________________ Date ____________

Claire went home after school and told her mother that 1 whole is the same as $\frac{2}{2}$ and $\frac{6}{6}$. Her mother asked why, but Claire couldn't explain. Use a number line and words to help Claire show and explain why $1 = \frac{2}{2} = \frac{6}{6}$.

Mr. Ramos wants to put a wire on the wall. He puts 9 nails equally spaced along the wire. Draw a number line representing the wire. Label it from 0 at the start of the wire to 1 at the end. Mark each fraction where Mr. Ramos puts each nail.

a. Build a number bond with unit fractions to 1 whole.

b. Write the fraction of the nail that is equivalent to $\frac{1}{2}$ of the wire.

Read **Draw** **Write**

Name ______________________________ Date ______________

1. Write the shaded fraction of each figure on the blank. Then, draw a line to match the equivalent fractions.

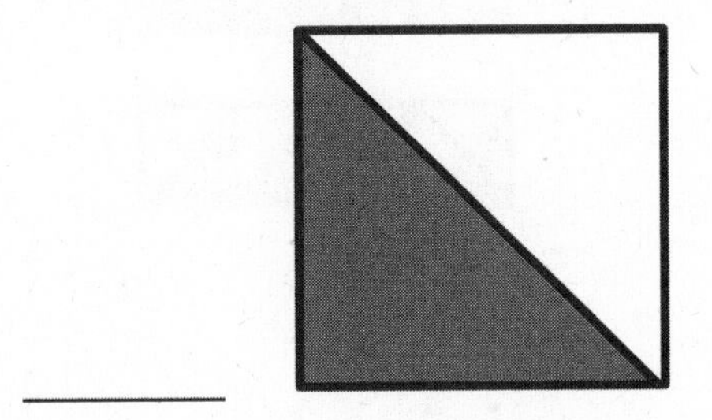

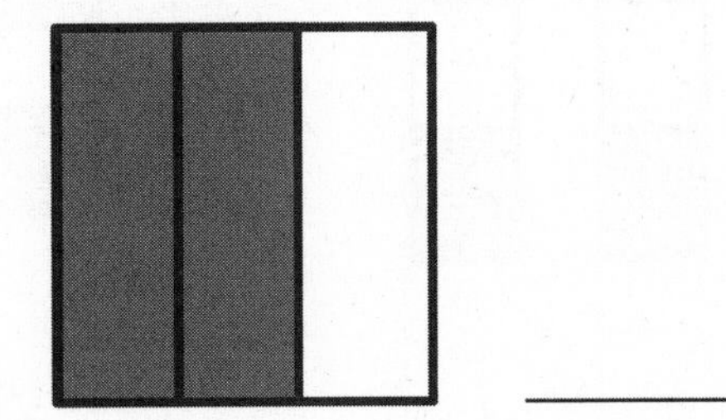

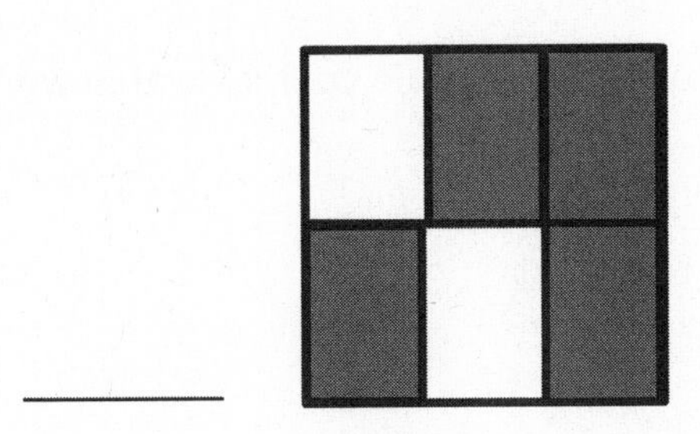

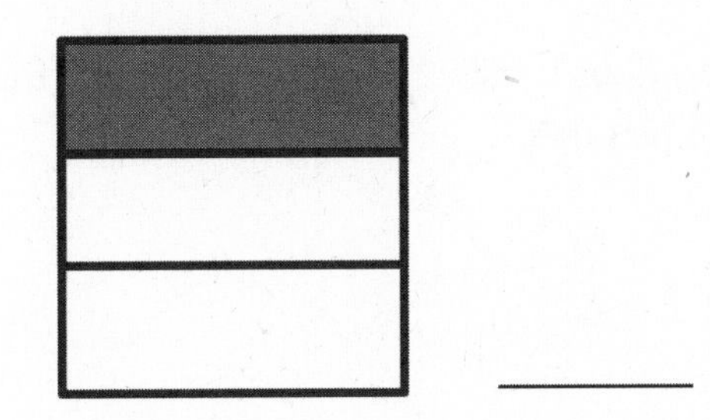

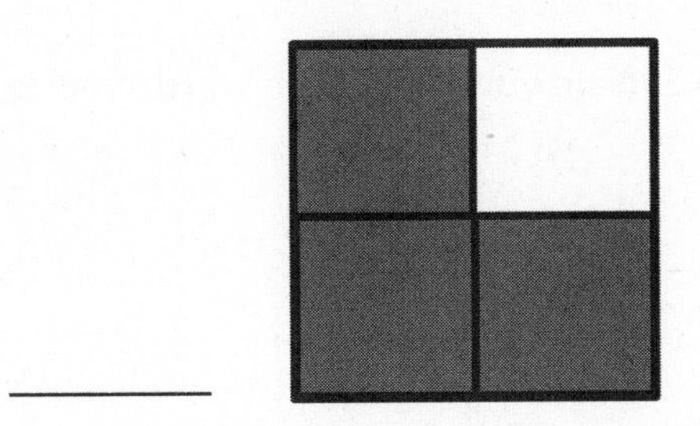

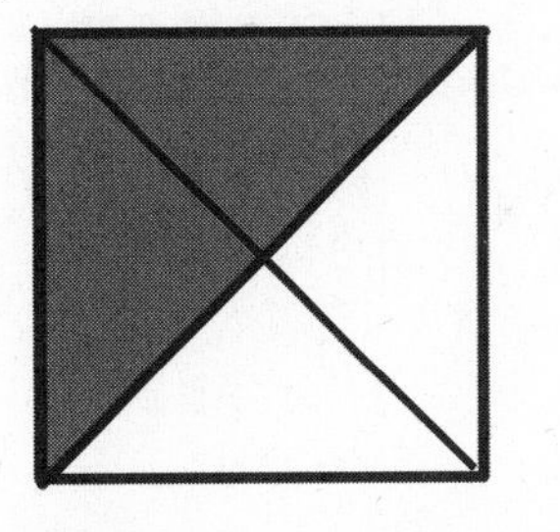

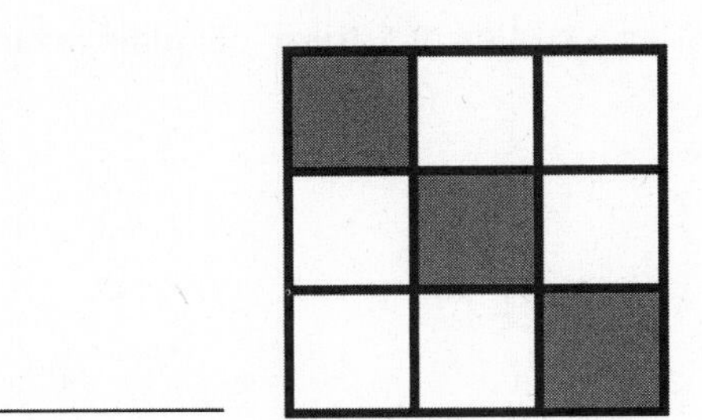

2. Write the missing parts of the fractions.

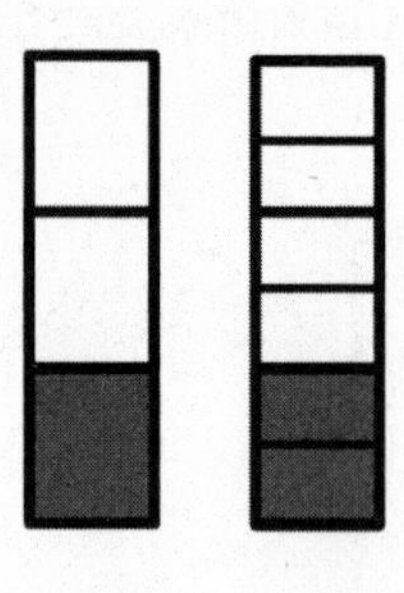

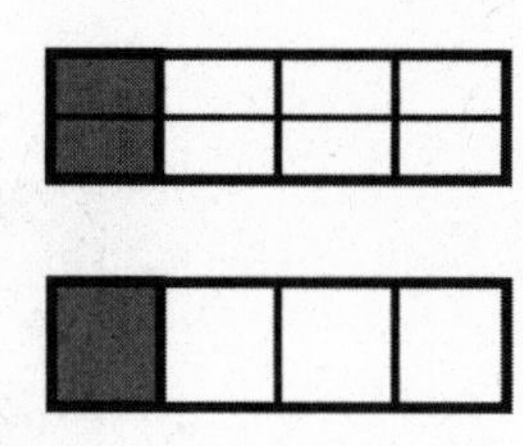

$\frac{1}{3} = \frac{\quad}{6}$ $\frac{2}{\quad} = \frac{1}{4}$ $\frac{4}{8} = \frac{8}{\quad}$

3. Why does it take 2 copies of $\frac{1}{8}$ to show the same amount as 1 copy of $\frac{1}{4}$? Explain your answer in words and pictures.

4. How many sixths does it take to make the same amount as $\frac{1}{3}$? Explain your answer in words and pictures.

5. Why does it take 10 copies of 1 sixth to make the same amount as 5 copies of 1 third? Explain your answer in words and pictures.

Name ______________________________ Date ______________

1. Draw and label two models that show equivalent fractions.

2. Draw a number line that proves your thinking about Problem 1.

Shannon stood at the end of a 100-meter long soccer field and kicked the ball to her teammate. She kicked it 20 meters. The commentator said she kicked it a quarter of the way down the field. Is that true? If not, what fraction should the commentator have said? Prove your answer by using a number line.

Read **Draw** **Write**

Name ______________________________ Date ________________

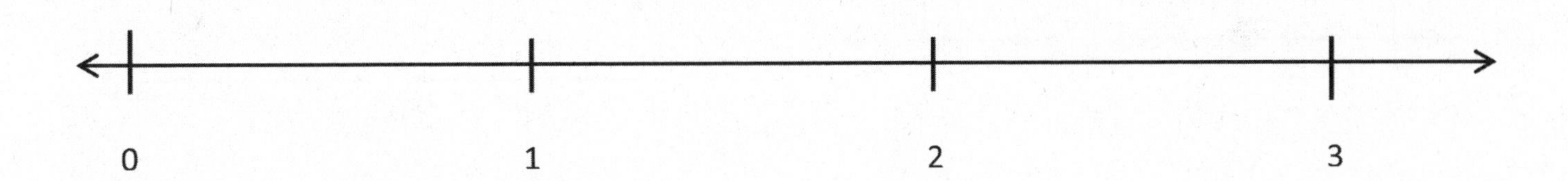

1. On the number line above, use a red colored pencil to divide each whole into fourths, and label each fraction above the line. Use a fraction strip to help you estimate, if necessary.

2. On the number line above, use a blue colored pencil to divide each whole into eighths, and label each fraction below the line. Refold your fraction strip from Problem 1 to help you estimate.

3. List the fractions that name the same place on the number line.

4. Using your number line to help, what red fraction and what blue fraction would be equal to $\frac{7}{2}$? Draw the part of the number line below that would include these fractions, and label it.

5. Write two different fractions for the dot on the number line. You may use halves, thirds, fourths, fifths, sixths, or eighths. Use fraction strips to help you, if necessary.

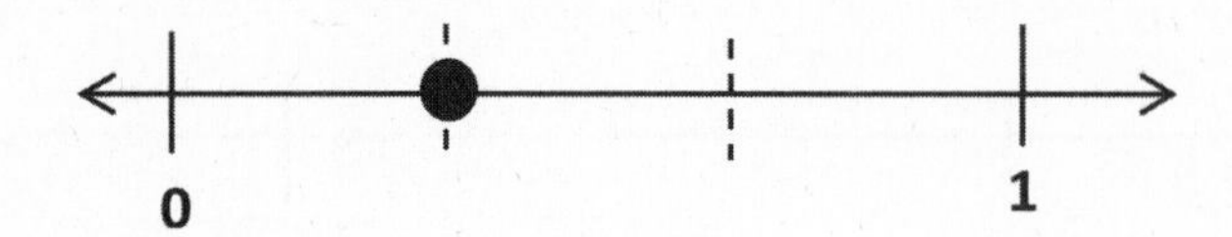

_____________ = _____________

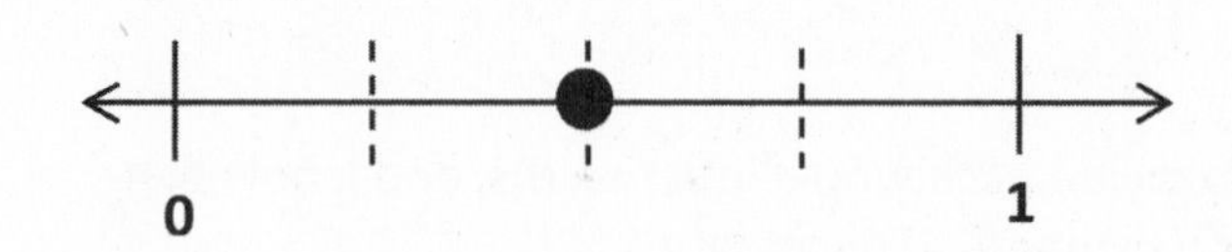

_____________ = _____________

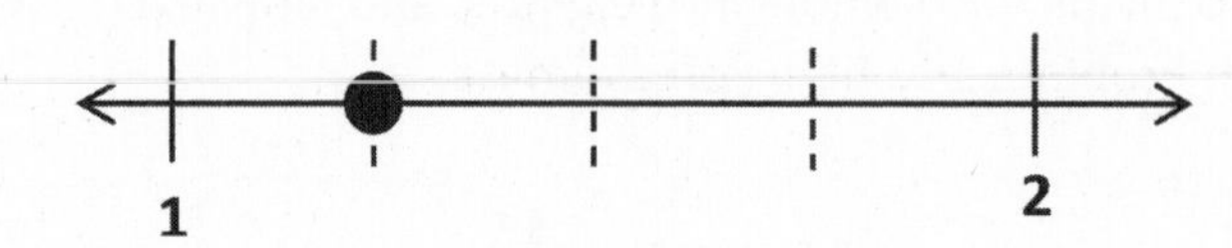

_____________ = _____________

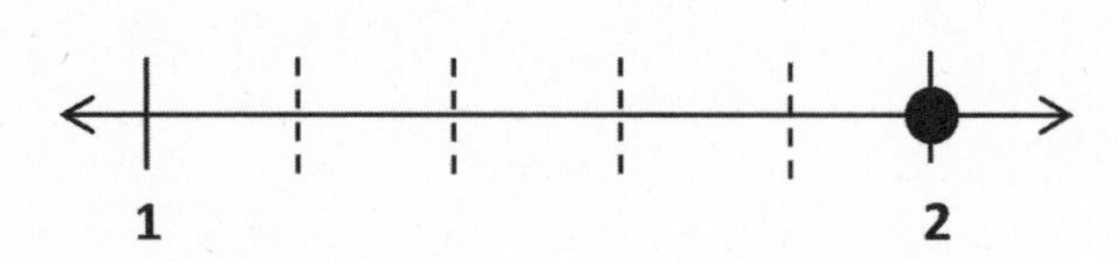

_____________ = _____________

6. Cameron and Terrance plan to run in the city race on Saturday. Cameron has decided that he will divide his race into 3 equal parts and will stop to rest after running 2 of them. Terrance divides his race into 6 equal parts and will stop and rest after running 2 of them. Will the boys rest at the same spot in the race? Why or why not? Draw a number line to explain your answer.

Name ____________________________________ Date ____________________

Henry and Maddie were in a pie-eating contest. The pies were cut either into thirds or sixths. Henry picked up a pie cut into sixths and ate $\frac{4}{6}$ of it in 1 minute. Maddie picked up a pie cut into thirds. What fraction of her pie does Maddie have to eat in 1 minute to tie with Henry? Draw a number line, and use words to explain your answer.

The zipper on Robert's jacket is 1 foot long. It breaks on the first day of winter. He can only zip it $\frac{8}{12}$ of the way before it gets stuck. Draw and label a number line to show how far Robert can zip his jacket.

a. Divide and label the number line in thirds. What fraction of the way can he zip his jacket in thirds?

b. What fraction of Robert's jacket is not zipped? Write your answer in twelfths and thirds.

Read **Draw** **Write**

Name ______________________________ Date ______________

1. Complete the number bond as indicated by the fractional unit. Partition the number line into the given fractional unit, and label the fractions. Rename 0 and 1 as fractions of the given unit. The first one is done for you.

Halves

1 — $\frac{1}{2}$, $\frac{1}{2}$

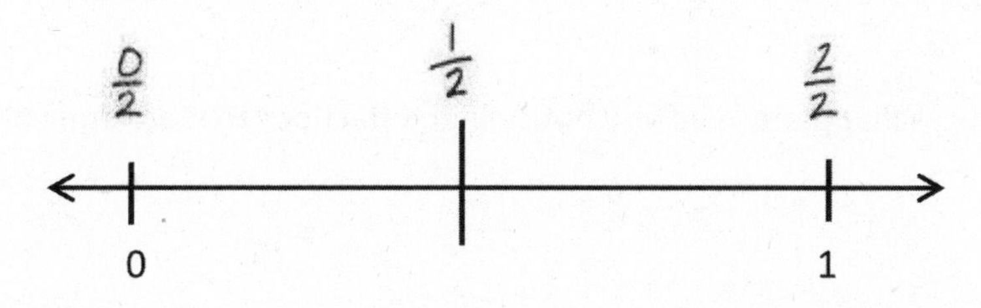

Thirds

1

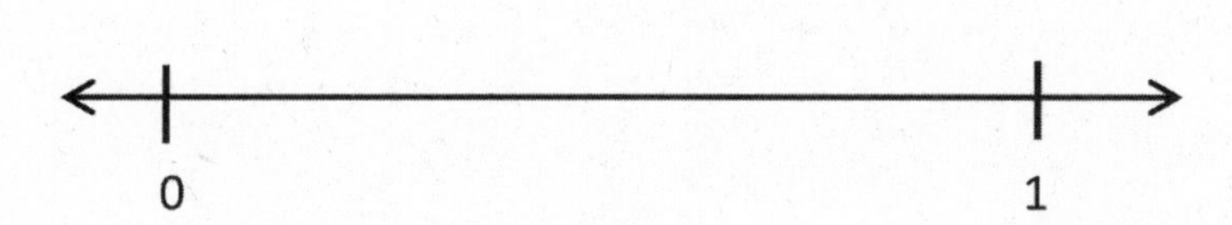

Fourths

1

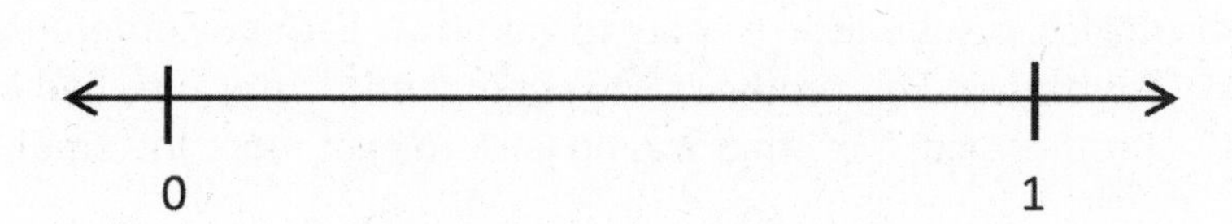

Fifths

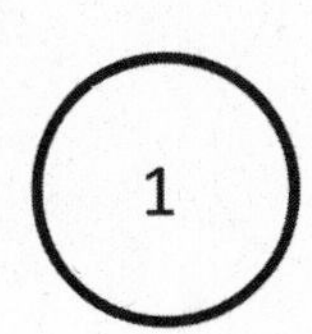

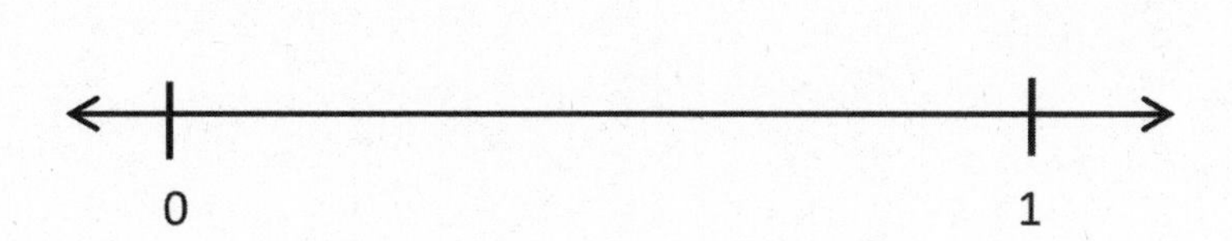

2. Circle all the fractions in Problem 1 that are equal to 1. Write them in a number sentence below.

$\frac{2}{2}$ = ________________ = ____________________ = ______________________

3. What pattern do you notice in the fractions that are equivalent to 1?

4. Taylor took his little brother to get pizza. Each boy ordered a small pizza. Taylor's pizza was cut in fourths, and his brother's was cut in thirds. After they had both eaten all of their pizza, Taylor's little brother said, "Hey that was no fair! You got more than me! You got 4 pieces, and I only got 3."

   Should Taylor's little brother be mad? What could you say to explain the situation to him? Use words, pictures, or a number line.

Name ______________________________ Date ______________

1. Complete the number bond as indicated by the fractional unit. Partition the number line into the given fractional unit, and label the fractions. Rename 0 and 1 as fractions of the given unit.

Fourths (1)

0 1

2. How many copies of $\frac{1}{4}$ does it take to make 1 whole? What's the fraction for 1 whole in this case? Use the number line or the number bond in Problem 1 to help you explain.

Lincoln drinks 1 eighth gallon of milk every morning.

a. How many days will it take Lincoln to drink 1 gallon of milk? Use a number line and words to explain your answer.

b. How many days will it take Lincoln to drink 2 gallons? Extend your number line to show 2 gallons, and use words to explain your answer.

Read **Draw** **Write**

Name ______________________________ Date ______________

1. Label the following models as a fraction inside the dotted box. The first one has been done for you.

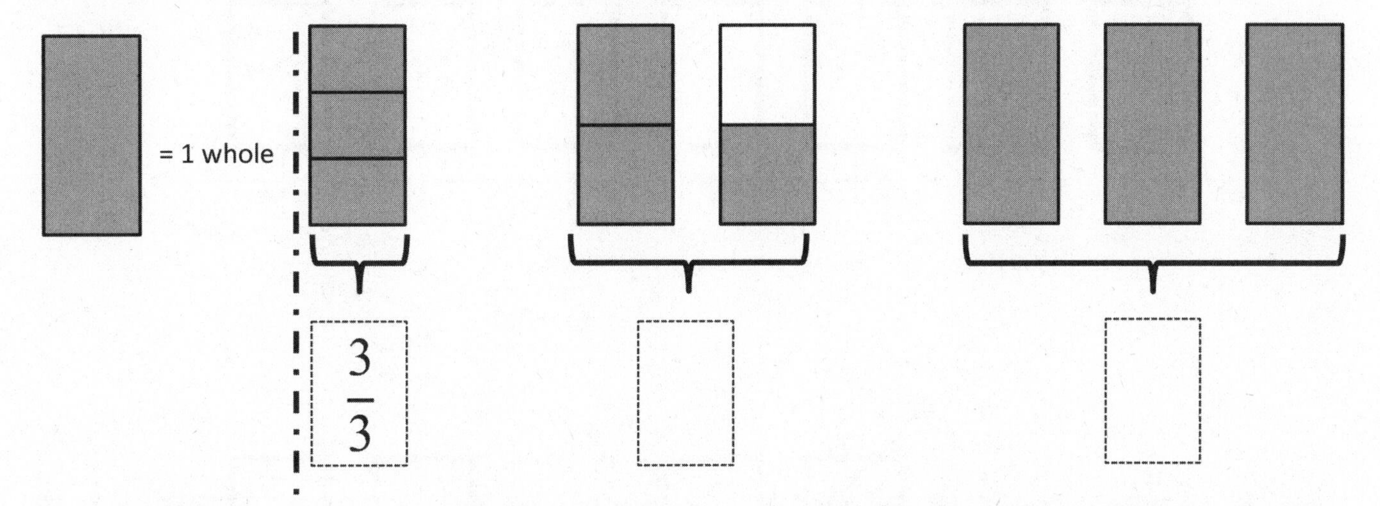

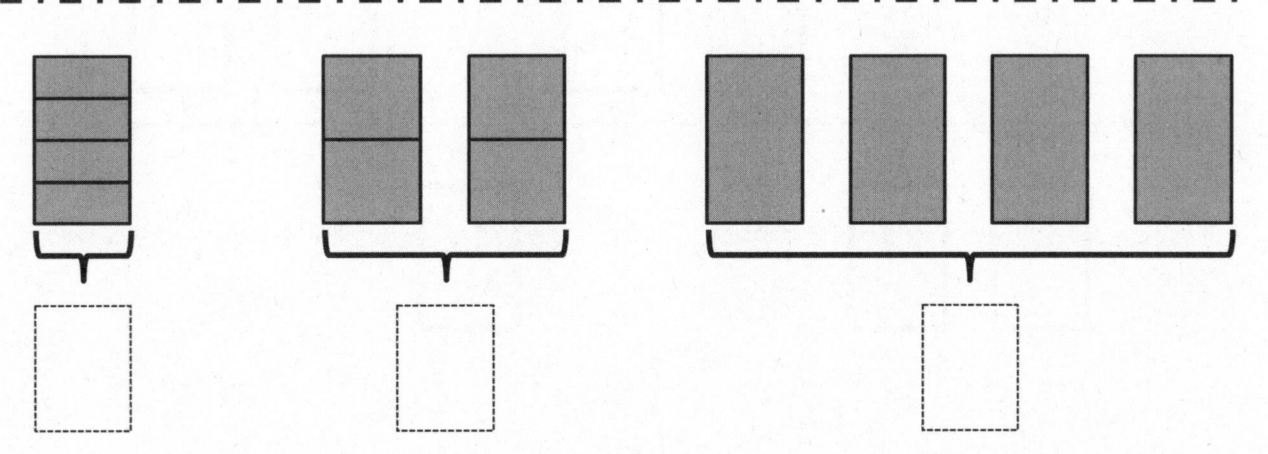

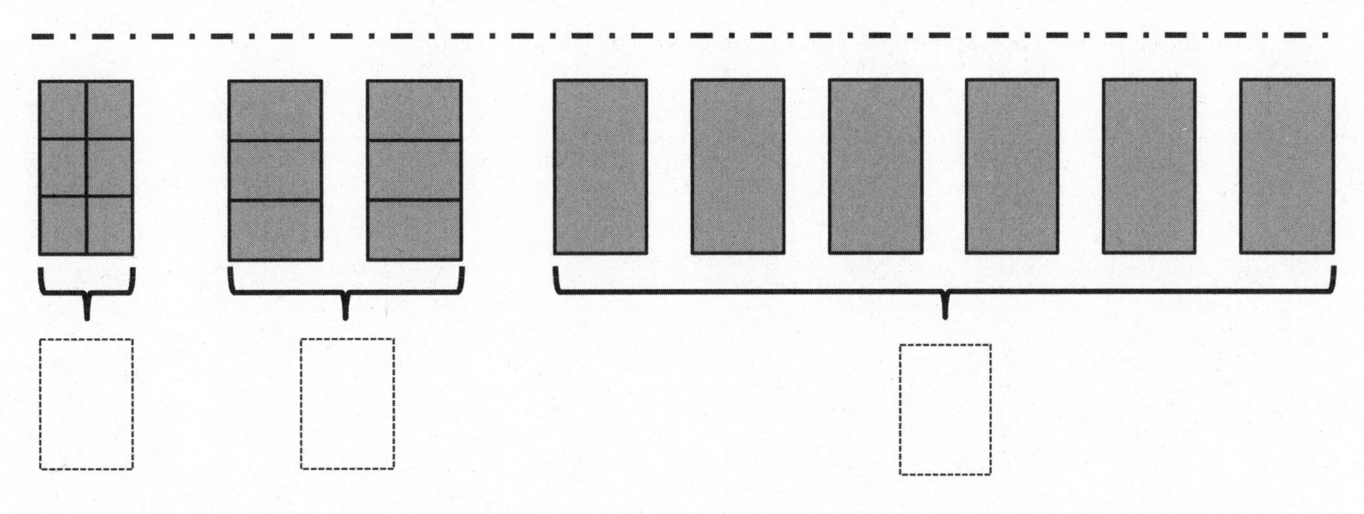

2. Fill in the missing whole numbers in the boxes below the number line. Rename the whole numbers as fractions in the boxes above the number line.

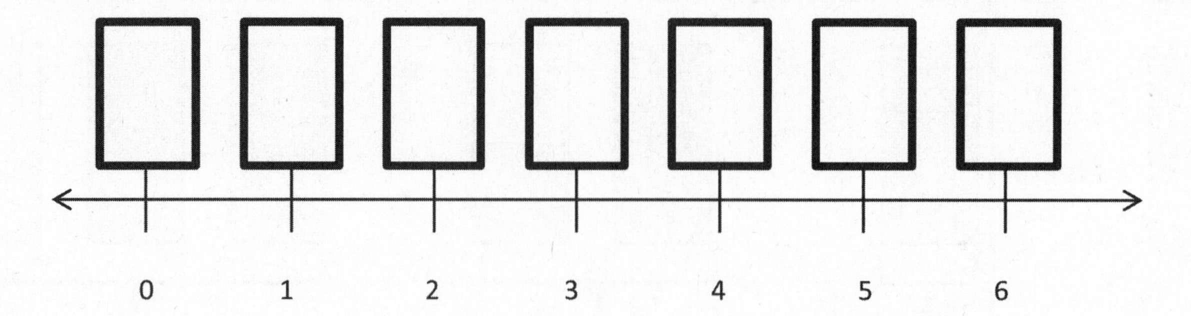

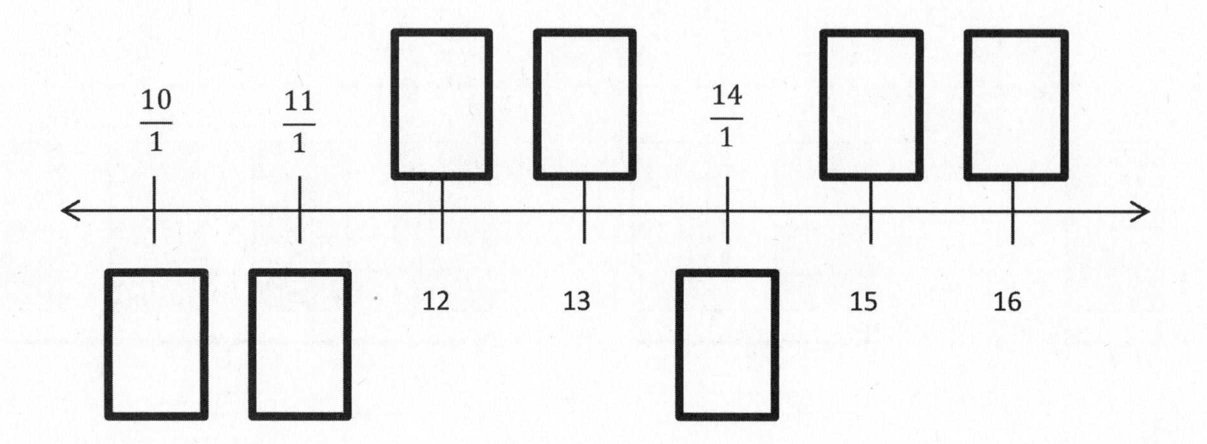

3. Explain the difference between these two fractions with words and pictures.

$$\frac{2}{1} \qquad \frac{2}{2}$$

Name ______________________________ Date ______________

1. Label the model as a fraction inside the box.

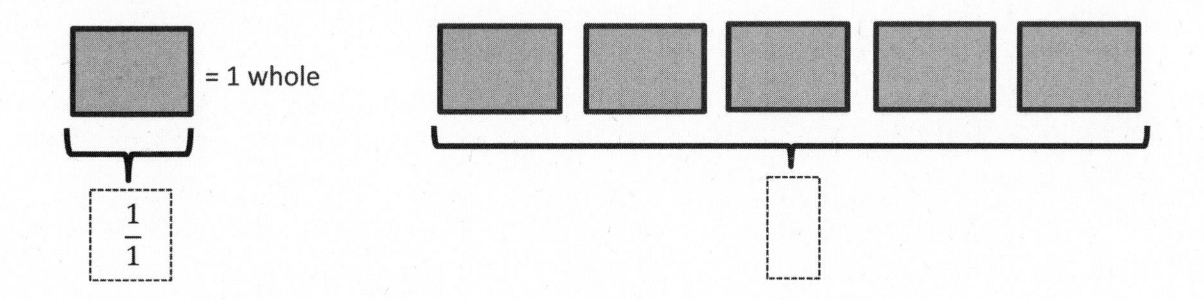

2. Partition the wholes into thirds. Rename the fraction for 3 wholes. Use the number line and words to explain your answer.

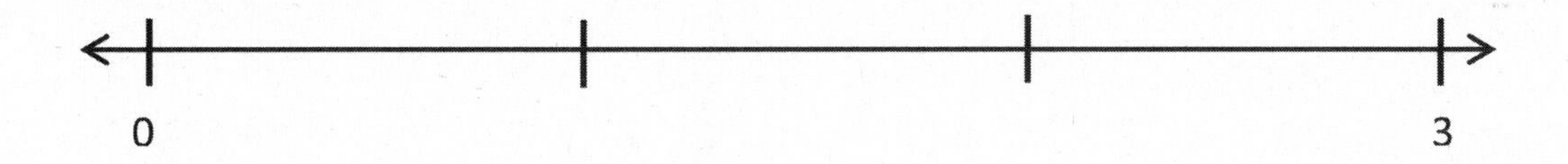

3 wholes

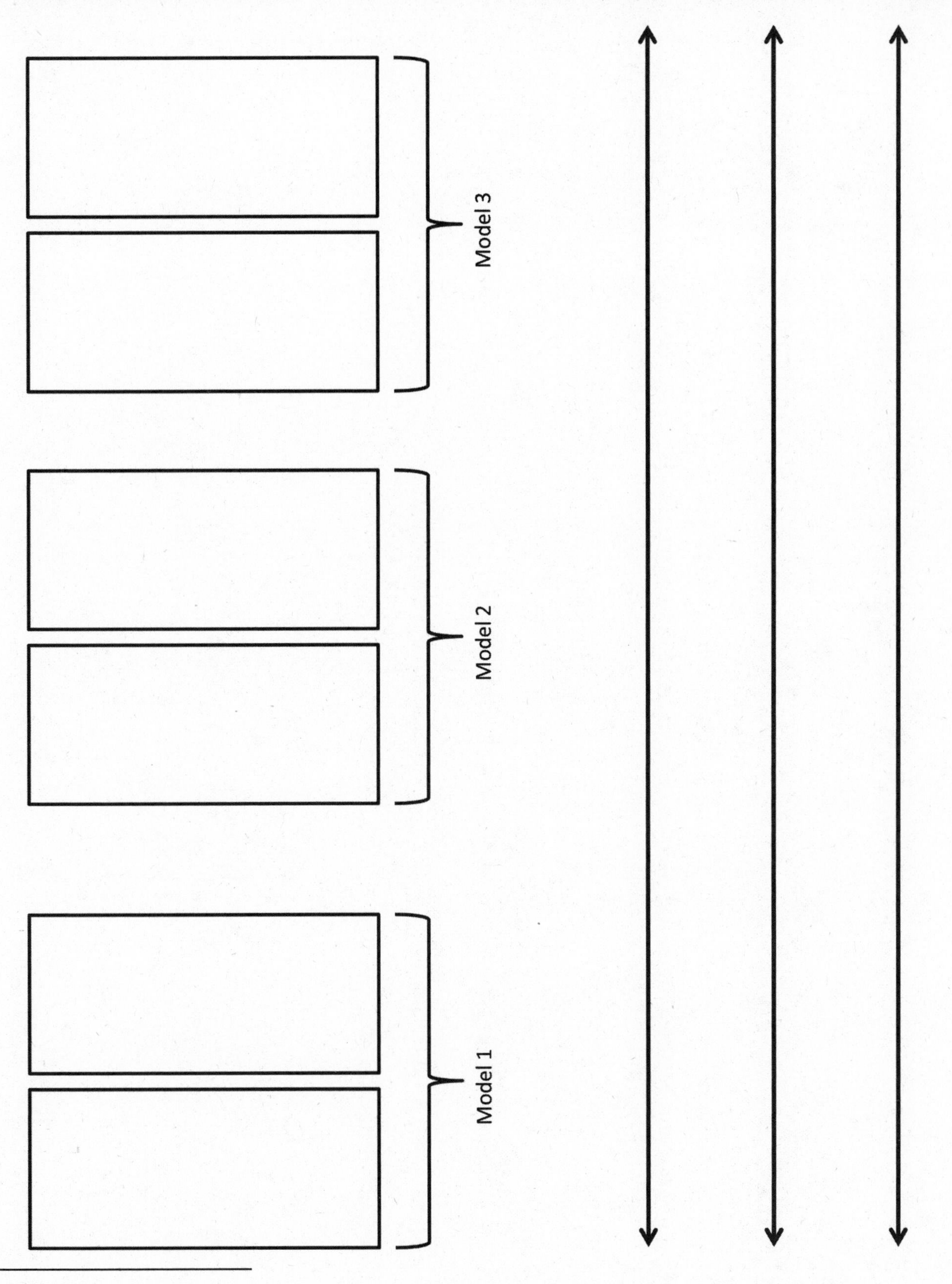

6 wholes

Antonio works on his project for 4 thirds hours. His mom tells him that he must spend another 2 thirds of an hour on it. Draw a number bond and number line with copies of thirds to show how long Antonio needs to work altogether. Write the amount of time Antonio needs to work altogether as a whole number.

Read **Draw** **Write**

Name ________________________________ Date ______________

1. Partition the number line to show the fractional units. Then, draw number bonds using copies of 1 whole for the circled whole numbers.

Halves

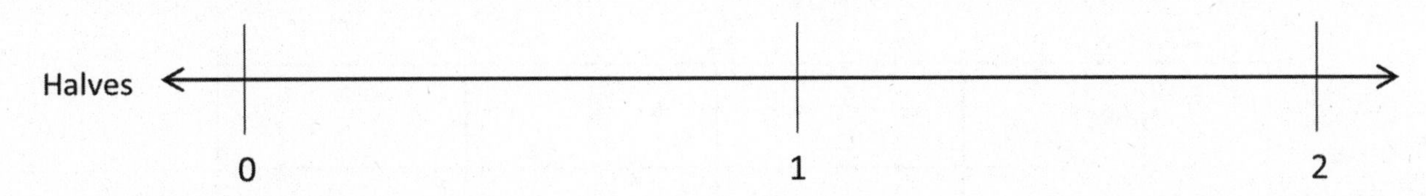

0 = _____ halves 1 = _____ halves 2 = _____ halves

$0 = \frac{\square}{2}$ $1 = \frac{\square}{2}$ $2 = \frac{4}{2}$

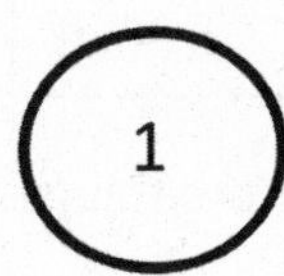

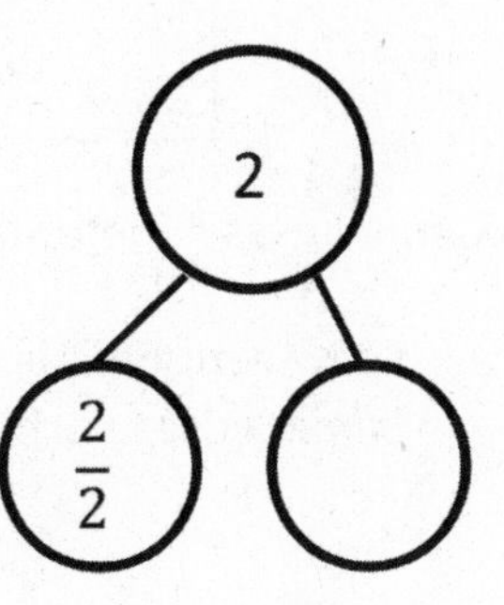

Thirds

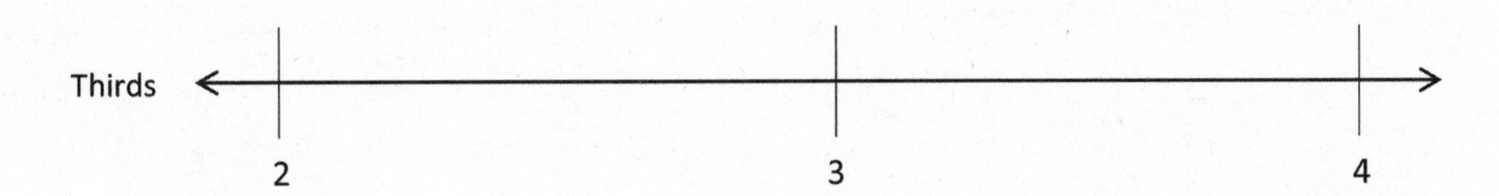

2 = _____ thirds 3 = _____ thirds 4 = _____ thirds

$2 = \frac{\square}{3}$ $3 = \frac{\square}{3}$ $4 = \frac{\square}{3}$

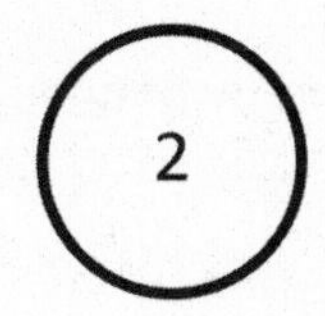

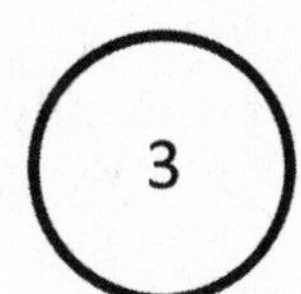

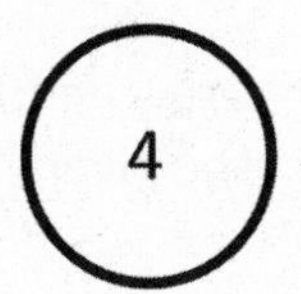

2. Write the fractions that name the whole numbers for each fractional unit. The first one has been done.

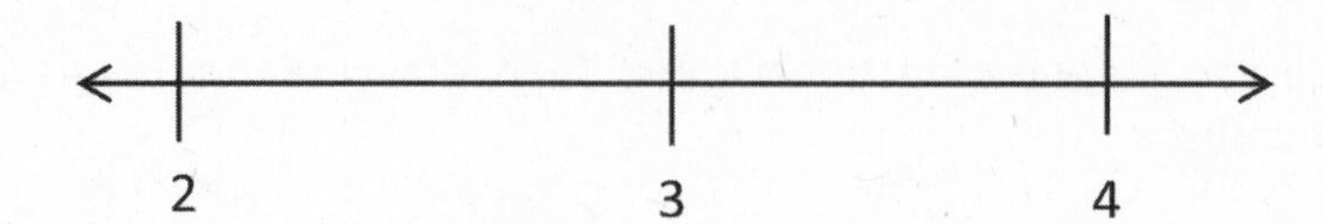

Halves	$\frac{4}{2}$	$\frac{6}{2}$	$\frac{8}{2}$
Thirds			
Fourths			
Sixths			

3. Sammy uses $\frac{1}{4}$ meter of wire each day to make things.

   a. Draw a number line to represent 1 meter of wire. Partition the number line to represent how much Sammy uses each day. How many days does the wire last?

   b. How many days will 3 meters of wire last?

4. Cindy feeds her dog $\frac{1}{3}$ pound of food each day.

   a. Draw a number line to represent 1 pound of food. Partition the number line to represent how much food she uses each day.

   b. Draw another number line to represent 4 pounds of food. After 3 days, how many pounds of food has she given her dog?

   c. After 6 days, how many pounds of food has she given her dog?

Name ______________________________ Date ______________

Irene has 2 yards of fabric.

a. Draw a number line to represent the total length of Irene's fabric.

b. Irene cuts her fabric into pieces of $\frac{1}{5}$ yard in length. Partition the number line to show her cuts.

c. How many $\frac{1}{5}$-yard pieces does she cut altogether? Use number bonds with copies of wholes to help you explain.

The branch of a tree is 2 meters long. Monica chops the branch for firewood. She cuts pieces that are $\frac{1}{6}$ meter long. Draw a number line to show the total length of the branch. Partition and label each of Monica's cuts.

a. How many pieces does Monica have altogether?

b. Write 2 equivalent fractions to describe the total length of Monica's branch.

Read **Draw** **Write**

Name ______________________________ Date ______________

1. Use the pictures to model equivalent fractions. Fill in the blanks, and answer the questions.

4 sixths is equal to 2 thirds.

$\frac{4}{6} = \frac{\square}{3}$

The whole stays the same.

What happened to the size of the equal parts when there were fewer equal parts?

They were bigger

What happened to the number of equal parts when the equal parts became larger?

There are fewer parts

1 half is equal to ______ eighths.

$\frac{1}{2} = \frac{\square}{8}$

The whole stays the same.

What happened to the size of the equal parts when there were more equal parts?

What happened to the number of equal parts when the equal parts became smaller?

2. 6 friends want to share 3 chocolate bars that are all the same size, which are represented by the 3 rectangles below. When the bars are unwrapped, the friends notice that the first chocolate bar is cut into 2 equal parts, the second is cut into 4 equal parts, and the third is cut into 6 equal parts. How can the 6 friends share the chocolate bars equally without breaking any of the pieces?

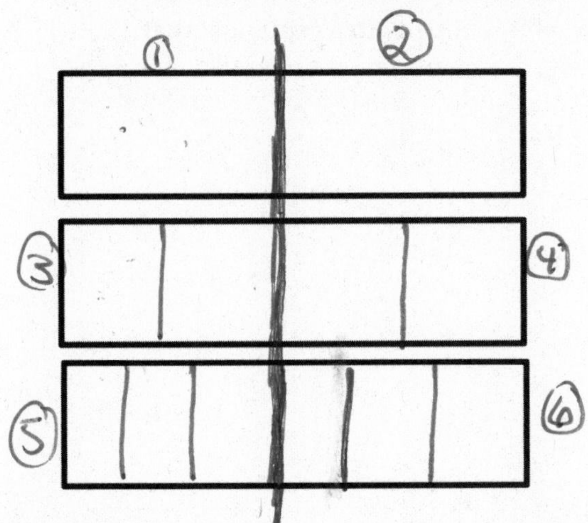

They each get one half.

$\frac{1}{2} = \frac{2}{4} = \frac{3}{6}$

3. When the whole is the same, why does it take 6 copies of 1 eighth to equal 3 copies of 1 fourth? Draw a model to support your answer.

4. When the whole is the same, how many sixths does it take to equal 1 third? Draw a model to support your answer.

5. You have a magic wand that doubles the number of equal parts but keeps the whole the same size. Use your magic wand. In the space below, draw to show what happens to a rectangle that is partitioned in fourths after you tap it with your wand. Use words and numbers to explain what happened.

Name ______________________________ Date ______________

1. Solve.

2 thirds is equal to _____ twelfths.

$$\frac{2}{3} = \frac{\quad}{12}$$

2. Draw and label two models that show fractions equivalent to those in Problem 1.

3. Use words to explain why the two fractions in Problem 1 are equal.

LaTonya has 2 equal-sized hotdogs. She cut the first one into thirds at lunch. Later, she cut the second hotdog to make double the number of pieces. Draw a model of LaTonya's hotdogs.

a. How many pieces is the second hotdog cut into?

b. If she wants to eat $\frac{2}{3}$ of the second hotdog, how many pieces should she eat?

Read **Draw** **Write**

Name ____________________________________ Date ____________________

Shade the models to compare the fractions. Circle the larger fraction for each problem.

1. 2 fifths

   2 thirds

2. 2 tenths

   2 eighths

3. 3 fourths

   3 eighths

4. 4 eighths

   4 sixths

5. 3 thirds

   3 sixths

6. After softball, Leslie and Kelly each buy a half-liter bottle of water. Leslie drinks 3 fourths of her water. Kelly drinks 3 fifths of her water. Who drinks the least amount of water? Draw a picture to support your answer.

7. Becky and Malory get matching piggy banks. Becky fills $\frac{2}{3}$ of her piggy bank with pennies. Malory fills $\frac{2}{4}$ of her piggy bank with pennies. Whose piggy bank has more pennies? Draw a picture to support your answer.

8. Heidi lines up her dolls in order from shortest to tallest. Doll A is $\frac{2}{4}$ foot tall, Doll B is $\frac{2}{6}$ foot tall, and Doll C is $\frac{2}{3}$ foot tall. Compare the heights of the dolls to show how Heidi puts them in order. Draw a picture to support your answer.

Name ______________________________ Date ______________

1. Shade the models to compare the fractions.

2 thirds

2 eighths

Which is larger, 2 thirds or 2 eighths? Why? Use words to explain.

2. Draw a model for each fraction. Circle the smaller fraction.

3 sevenths

3 fourths

Catherine and Diana buy matching scrapbooks. Catherine decorates $\frac{5}{9}$ of the pages in her book. Diana decorates $\frac{5}{6}$ of the pages in her book. Who has decorated more pages of her scrapbook? Draw a picture to support your answer.

____

____

____

____

Read **Draw** **Write**

Name ____________________________________ Date ________________

Label each shaded fraction. Use >, <, or = to compare. The first one has been done for you.

1.

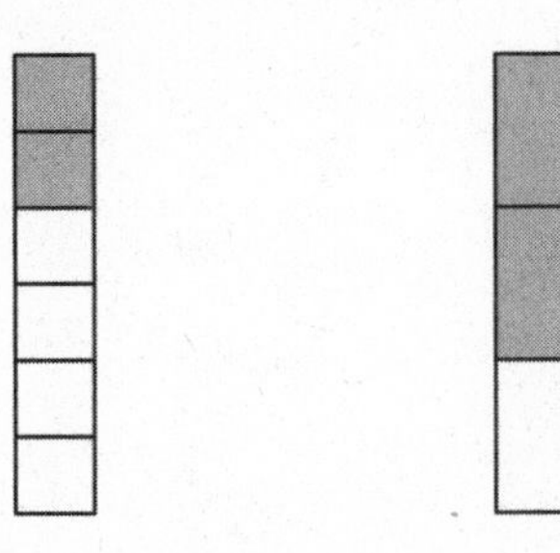

$\frac{2}{6}$ < $\frac{2}{3}$

2.

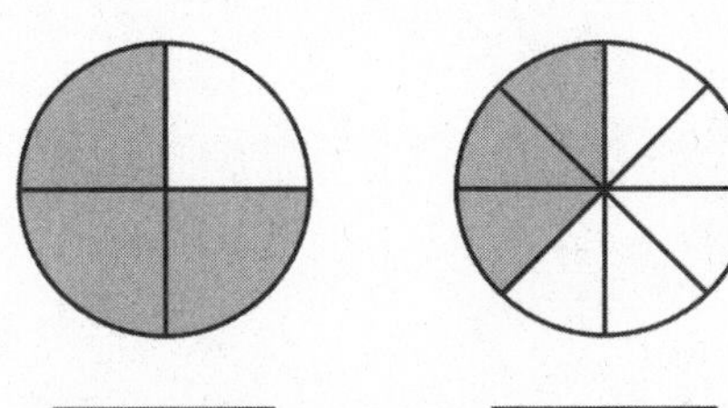

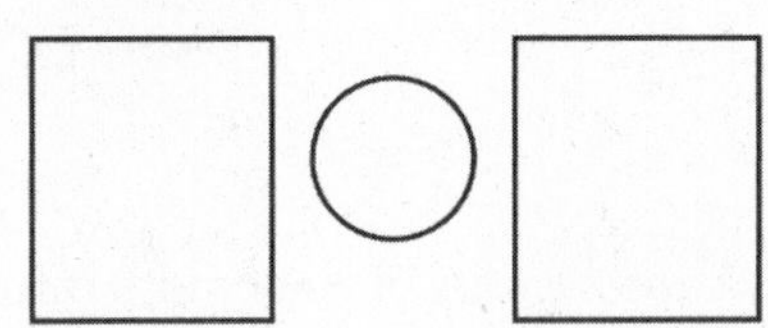

3.

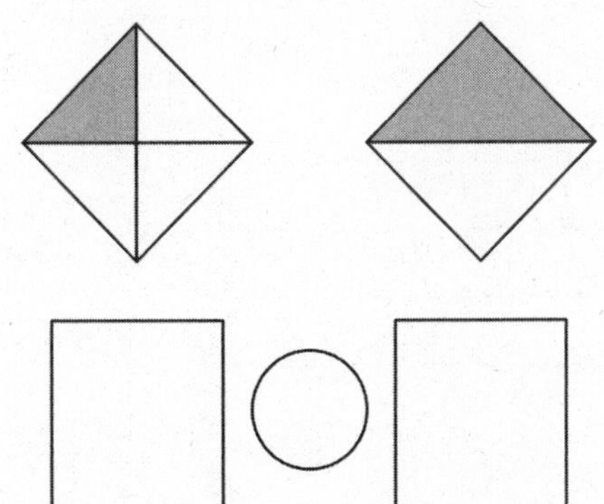

4.

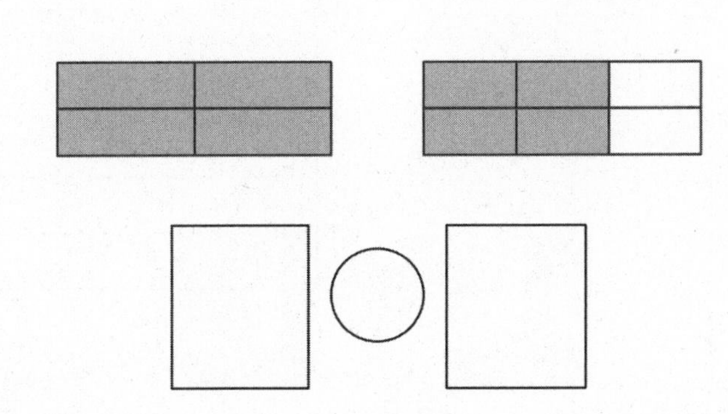

5. Partition each number line into the units labeled on the left. Then, use the number lines to compare the fractions.

halves

0 1

fourths

0 1

eighths

0 1

a. $\frac{3}{8}$ ◯ $\frac{3}{4}$

b. $\frac{4}{4}$ ◯ $\frac{4}{8}$

c. $\frac{2}{4}$ ◯ $\frac{2}{8}$

Draw your own model to compare the following fractions.

6. $\frac{3}{10}$ ◯ $\frac{3}{5}$

7. $\frac{2}{6}$ ◯ $\frac{2}{8}$

8. John ran 2 thirds of a kilometer after school. Nicholas ran 2 fifths of a kilometer after school. Who ran the shorter distance? Use the model below to support your answer. Be sure to label 1 whole as 1 kilometer.

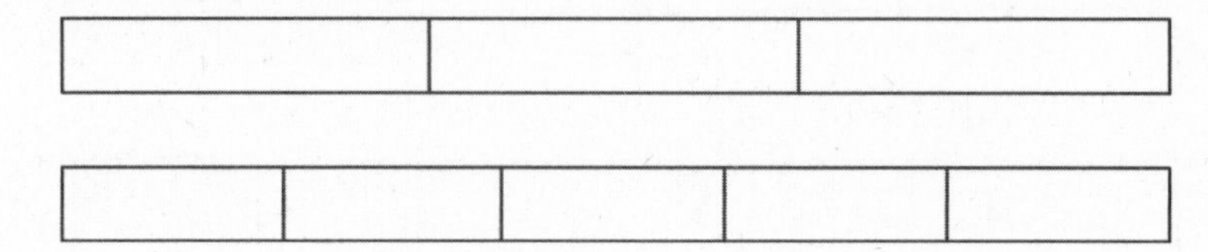

9. Erica ate 2 ninths of a licorice stick. Robbie ate 2 fifths of an identical licorice stick. Who ate more? Use the model below to support your answer.

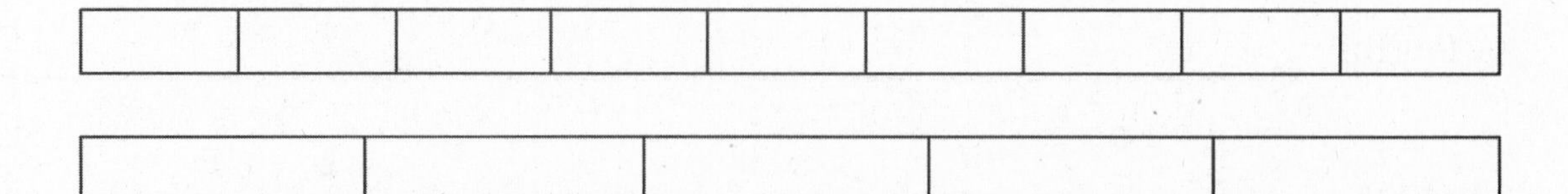

Name ______________________________ Date ______________

1. Complete the number sentence by writing >, <, or =.

   $\frac{3}{5}$ ________ $\frac{3}{9}$

2. Draw 2 number lines with endpoints 0 and 1 to show each fraction in Problem 1. Use the number lines to explain how you know your comparison in Problem 1 is correct.

lined paper

Grade 3
Module 6

Damien folds a paper strip into 6 equal parts. He shades 5 of the equal parts and then cuts off 2 shaded parts. Explain your thinking about what fraction is unshaded.

Read **Draw** **Write**

Name ________________________ Date ____________

1. "What is your favorite color?" Survey the class to complete the tally chart below.

Favorite Colors	
Color	**Number of Students**
Green	||| 3
Yellow	|| 2
Red	~~||||~~ 5
Blue	~~||||~~ ~~||||~~ ~~||||~~ ~~||||~~ 20
Orange	||| 3

2. Use the tally chart to answer the following questions.

   a. How many students chose orange as their favorite color? 3

   b. How many students chose yellow as their favorite color? 2

   c. Which color did students choose the most? How many students chose it? blue, 20

   d. Which color did students choose the least? How many students chose it? yellow, 2

   e. What is the difference between the number of students in parts (c) and (d)? Write a number sentence to show your thinking. $20 - 2 = 18$

   f. Write an equation to show the total number of students surveyed on this chart.

   $3 + 2 + 5 + 20 + 3 = 33$

3. Use the tally chart in Problem 1 to complete the picture graphs below.

a.

Favorite Colors				
Green	Yellow	Red	Blue	Orange
Each ♥ represents 1 student.				

b.

Favorite Colors				
Green	Yellow	Red	Blue	Orange
Each ♥ represents 2 students.				

4. Use the picture graph in Problem 3(b) to answer the following questions.

   a. What does each ♡ represent?

   b. Draw a picture and write a number sentence to show how to represent 3 students in your picture graph.

   c. How many students does ♡ ♡ ♡ ♡ ♡ ♡ ♡ represent? Write a number sentence to show how you know.

   d. How many more ♡ did you draw for the color that students chose the most than for the color that students chose the least? Write a number sentence to show the difference between the number of votes for the color that students chose the most and the color that students chose the least.

Name ______________________________ Date ______________

The picture graph below shows data from a survey of students' favorite sports.

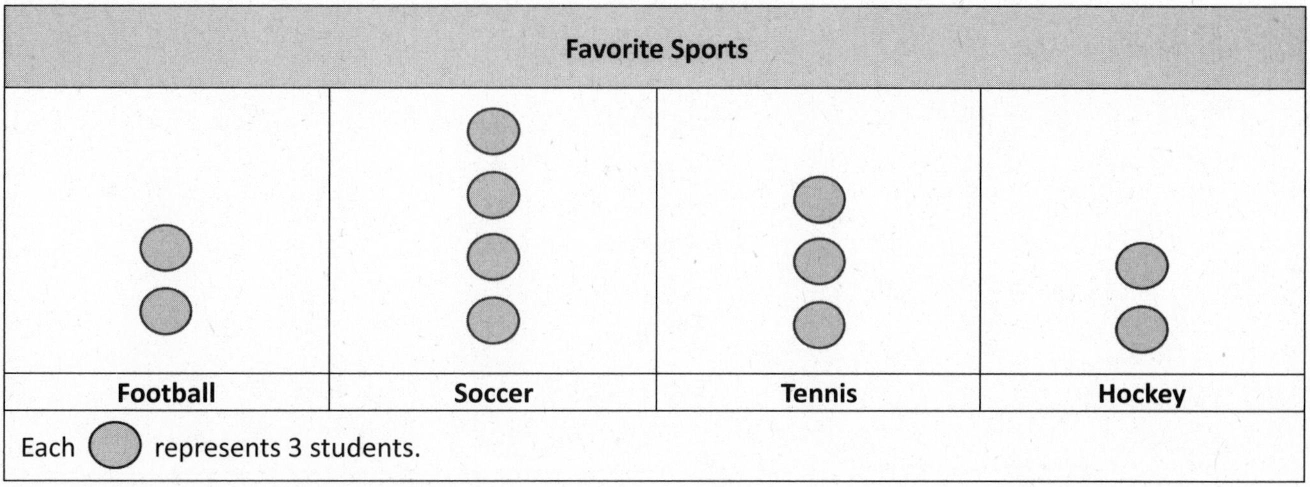

Favorite Sports			
● ●	● ● ● ●	● ● ●	● ●
Football	**Soccer**	**Tennis**	**Hockey**
Each ● represents 3 students.			

a. The same number of students picked __________ and __________ as their favorite sport.

b. How many students picked tennis as their favorite sport?

c. How many more students picked soccer than tennis? Use a number sentence to show your thinking.

d. How many total students were surveyed?

Reisha played in three basketball games. She scored 12 points in Game 1, 8 points in Game 2, and 16 points in Game 3. Each basket that she made was worth 2 points. She uses tape diagrams with a unit size of 2 to represent the points she scored in each game. How many total units of 2 does it take to represent the points she scored in all three games?

Read **Draw** **Write**

Name ______________________________ Date ________________

1. Find the total number of stamps each student has. Draw tape diagrams with a unit size of 4 to show the number of stamps each student has. The first one has been done for you.

Dana

Tanisha

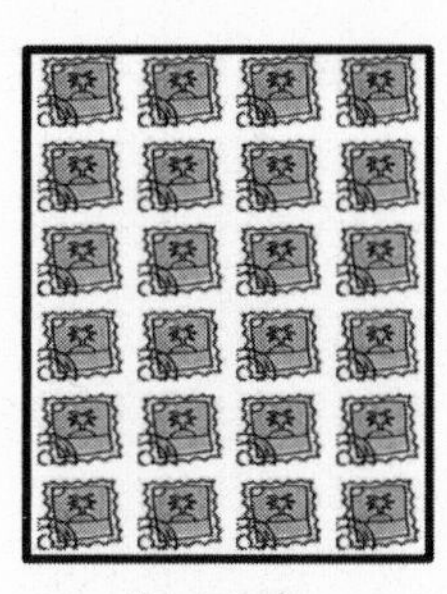

Raquel

Anna

Each stamp represents 1 stamp.

Dana:

4	4	4	4

Tanisha:

Raquel:

Anna:

2. Explain how you can create vertical tape diagrams to show this data.

3. Complete the vertical tape diagrams below using the data from Problem 1.

a.

4
4
4
4

Dana Tanisha Raquel Anna

b.

8
8

Dana Tanisha Raquel Anna

c. What is a good title for the vertical tape diagrams?

d. How many total units of 4 are in the vertical tape diagrams in Problem 3(a)?

e. How many total units of 8 are in the vertical tape diagrams in Problem 3(b)?

f. Compare your answers to parts (d) and (e). Why does the number of units change?

g. Mattaeus looks at the vertical tape diagrams in Problem 3(b) and finds the total number of Anna's and Raquel's stamps by writing the equation $7 \times 8 = 56$. Explain his thinking.

Name ______________________________ Date ______________

The chart below shows a survey of the book club's favorite type of book.

Book Club's Favorite Type of Book	
Type of Book	**Number of Votes**
Mystery	12
Biography	16
Fantasy	20
Science Fiction	8

a. Draw tape diagrams with a unit size of 4 to represent the book club's favorite type of book.

b. Use your tape diagrams to draw vertical tape diagrams that represent the data.

The vertical tape diagrams show the number of fish in Sal's Pet Store.

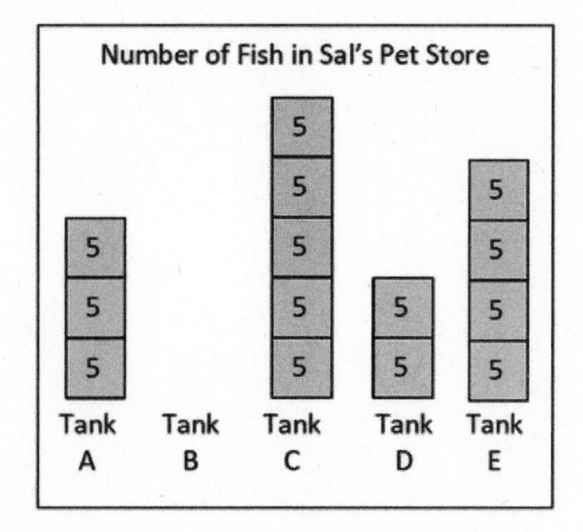

a. Find the total number of fish in Tank C. Show your work.

b. Tank B has a total of 30 fish. Draw the tape diagram for Tank B.

Read **Draw** **Write**

c. How many more fish are in Tank B than in Tanks A and D combined?

Read **Draw** **Write**

Name ______________________________ Date ______________

1. This table shows the number of students in each class.

Number of Students in Each Class	
Class	Number of Students
Baking	9
Sports	16
Chorus	13
Drama	18

Use the table to color the bar graph. The first one has been done for you.

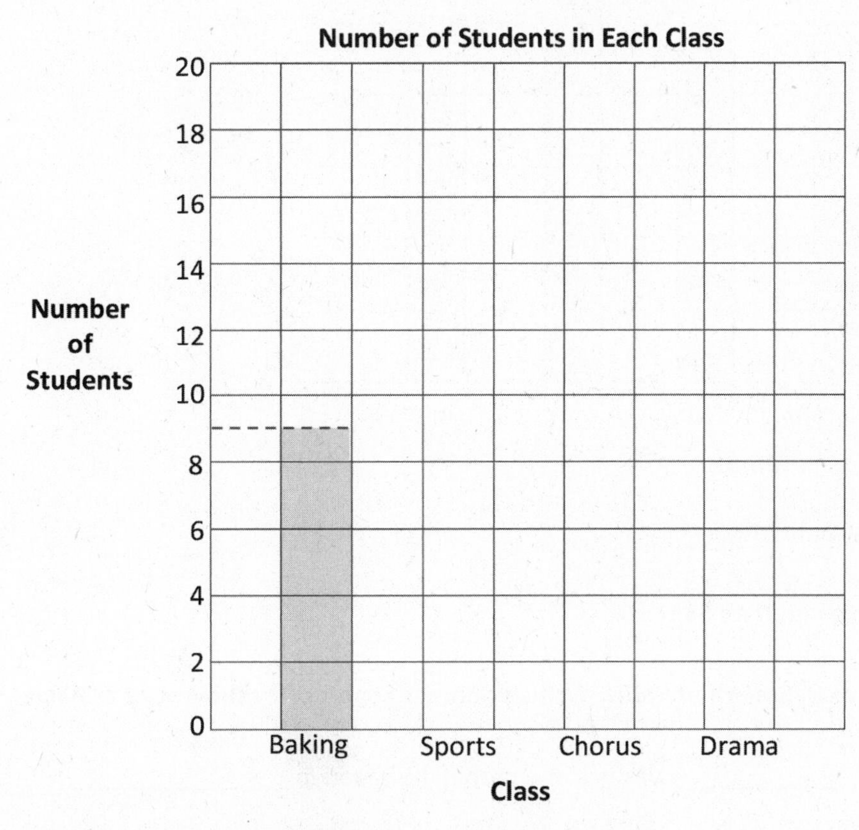

a. What is the value of each square in the bar graph?

b. Write a number sentence to find how many total students are enrolled in classes.

c. How many fewer students are in sports than in chorus and baking combined? Write a number sentence to show your thinking.

2. This bar graph shows Kyle's savings from February to June. Use a straightedge to help you read the graph.

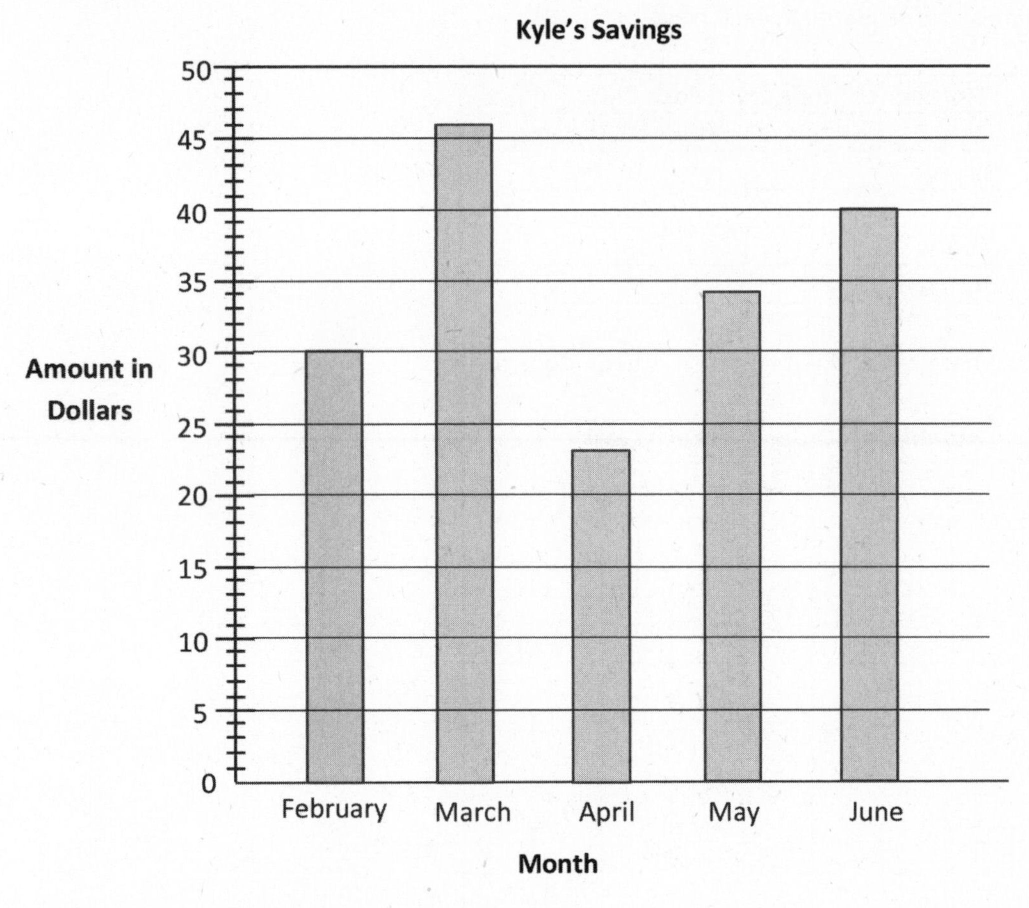

a. How much money did Kyle save in May?

b. In which months did Kyle save less than $35?

c. How much more did Kyle save in June than April? Write a number sentence to show your thinking.

d. The money Kyle saved in _______________ was half the money he saved in _______________.

3. Complete the table below to show the same data given in the bar graph in Problem 2.

Months	February				
Amount Saved in Dollars					

This bar graph shows the number of minutes Charlotte read from Monday through Friday.

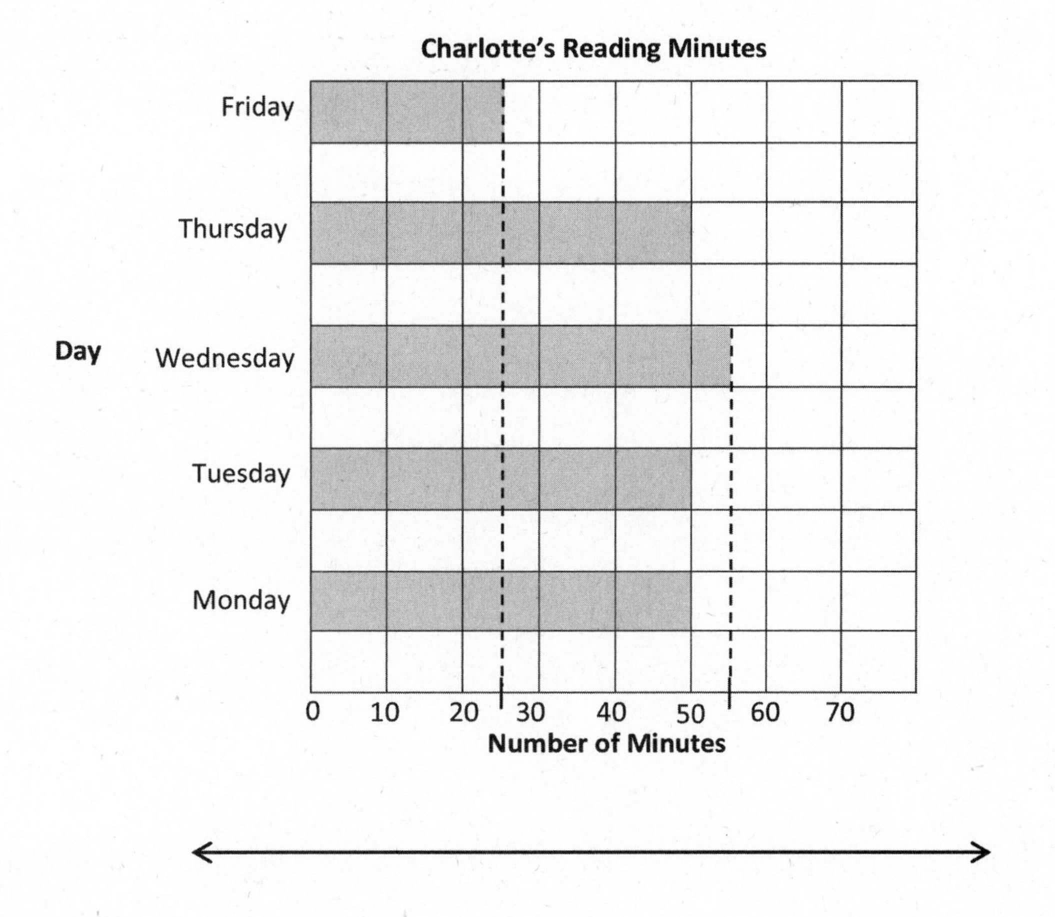

4. Use the graph's lines as a ruler to draw in the intervals on the number line shown above. Then plot and label a point for each day on the number line.

5. Use the graph or number line to answer the following questions.

   a. On which days did Charlotte read for the same number of minutes? How many minutes did Charlotte read on these days?

   b. How many more minutes did Charlotte read on Wednesday than on Friday?

Name ______________________________ Date ______________

The bar graph below shows the students' favorite ice cream flavors.

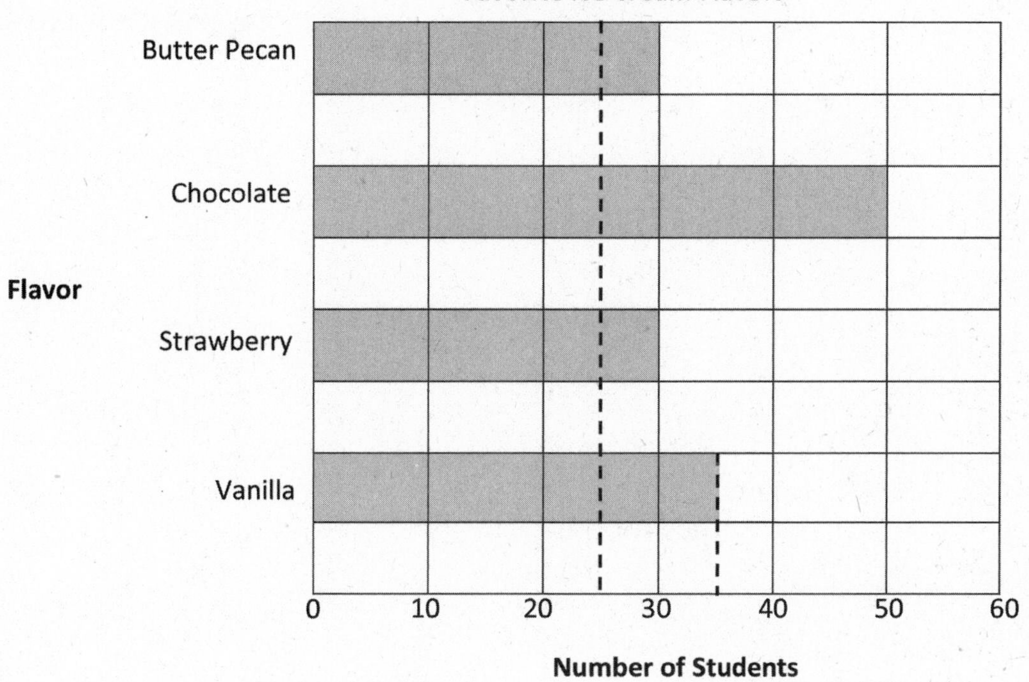

a. Use the graph's lines as a ruler to draw intervals on the number line shown above. Then plot and label a point for each flavor on the number line.

b. Write a number sentence to show the total number of students who voted for butter pecan, vanilla, and chocolate.

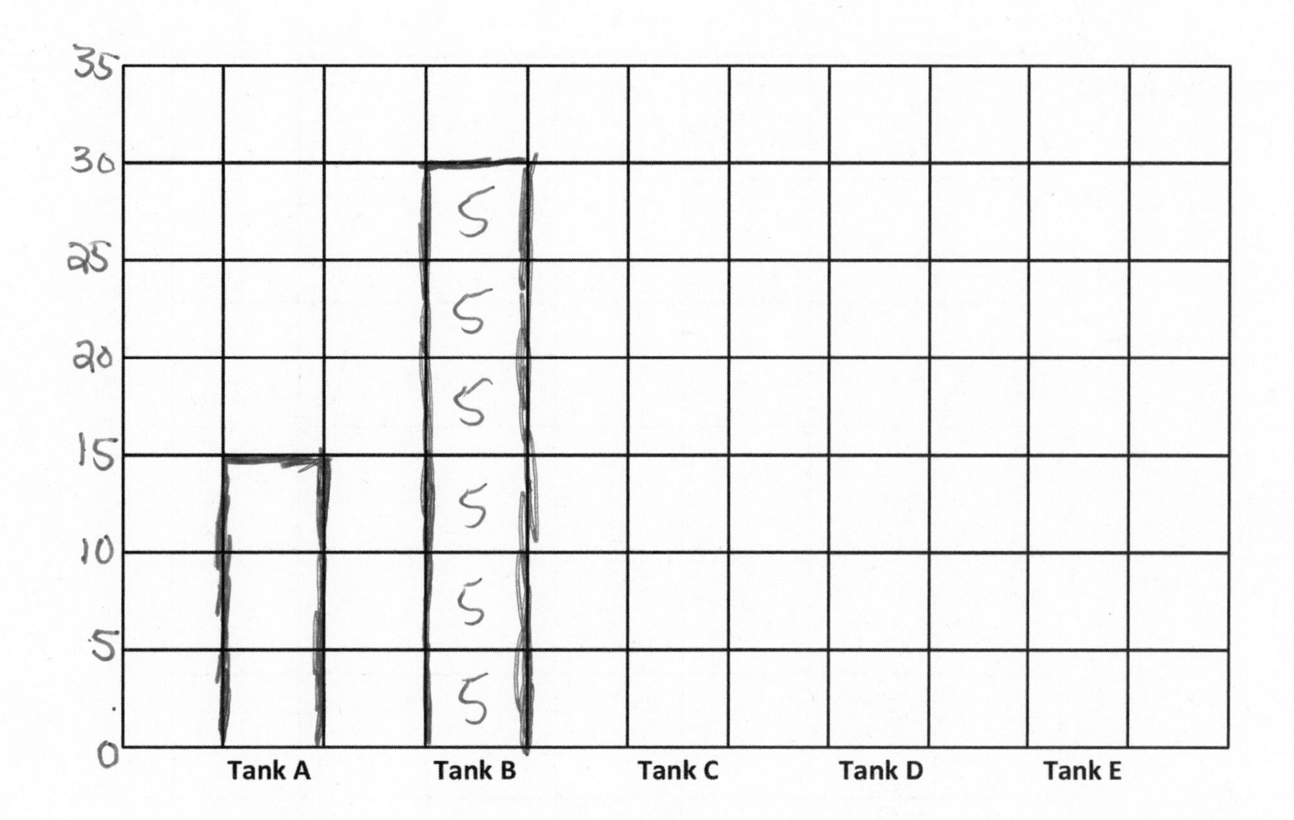

graph A

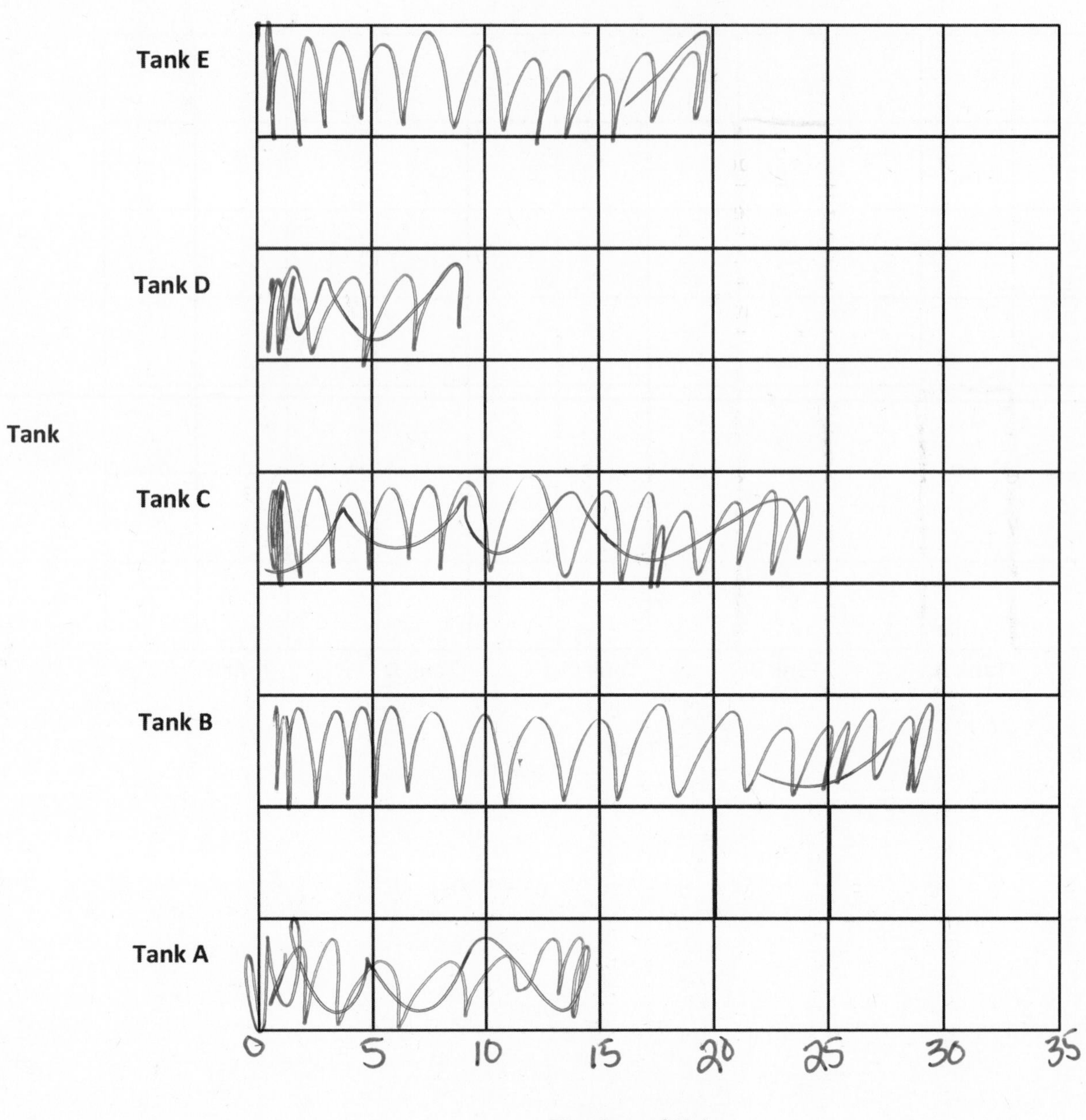

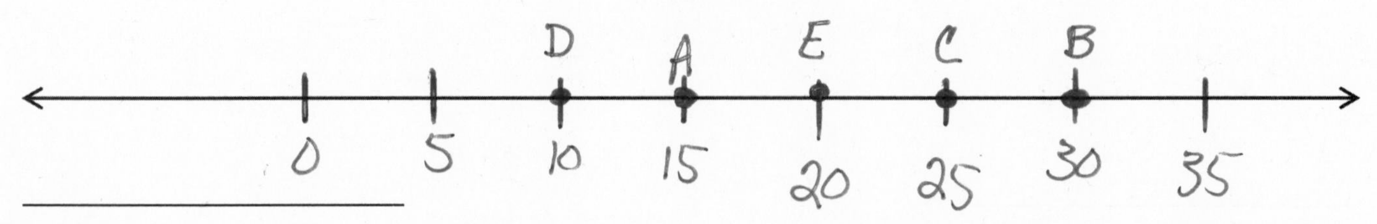

graph B

EUREKA MATH

The following chart shows the number of times an insect's wings vibrate each second. Use the following clues to complete the unknowns in the chart.

Wing Vibrations of Insects	
Insect	**Number of Wing Vibrations Each Second**
Honeybee	350
Beetle	b
Fly	550
Mosquito	m

a. The beetle's number of wing vibrations is the same as the difference between the fly's and honeybee's.

b. The mosquito's number of wing vibrations is the same as 50 less than the beetle's and fly's combined.

Read **Draw** **Write**

Name ____________________ Date __________

1. The chart below shows the number of magazines sold by each student.

Student	Ben	Rachel	Jeff	Stanley	Debbie
Magazines Sold	300	250	100	450	600

a. Use the chart to draw a bar graph below. Create an appropriate scale for the graph.

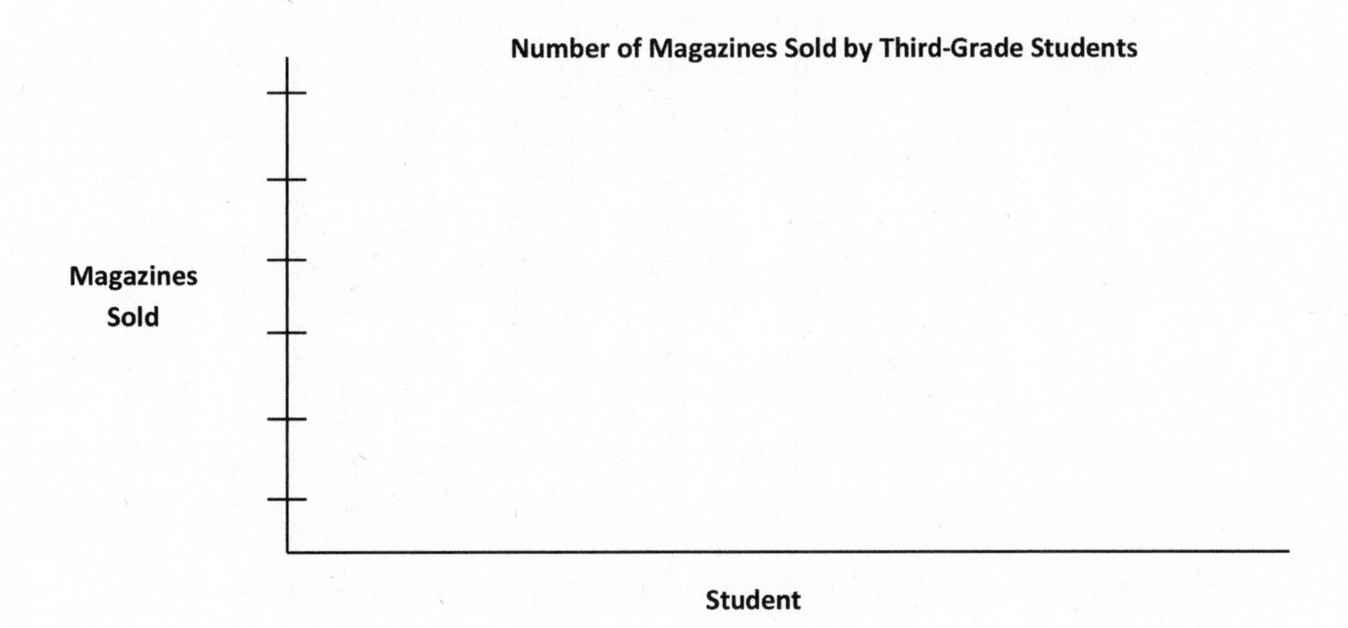

b. Explain why you chose the scale for the graph.

c. How many fewer magazines did Debbie sell than Ben and Stanley combined?

d. How many more magazines did Debbie and Jeff sell than Ben and Rachel?

2. The bar graph shows the number of visitors to a carnival from Monday through Friday.

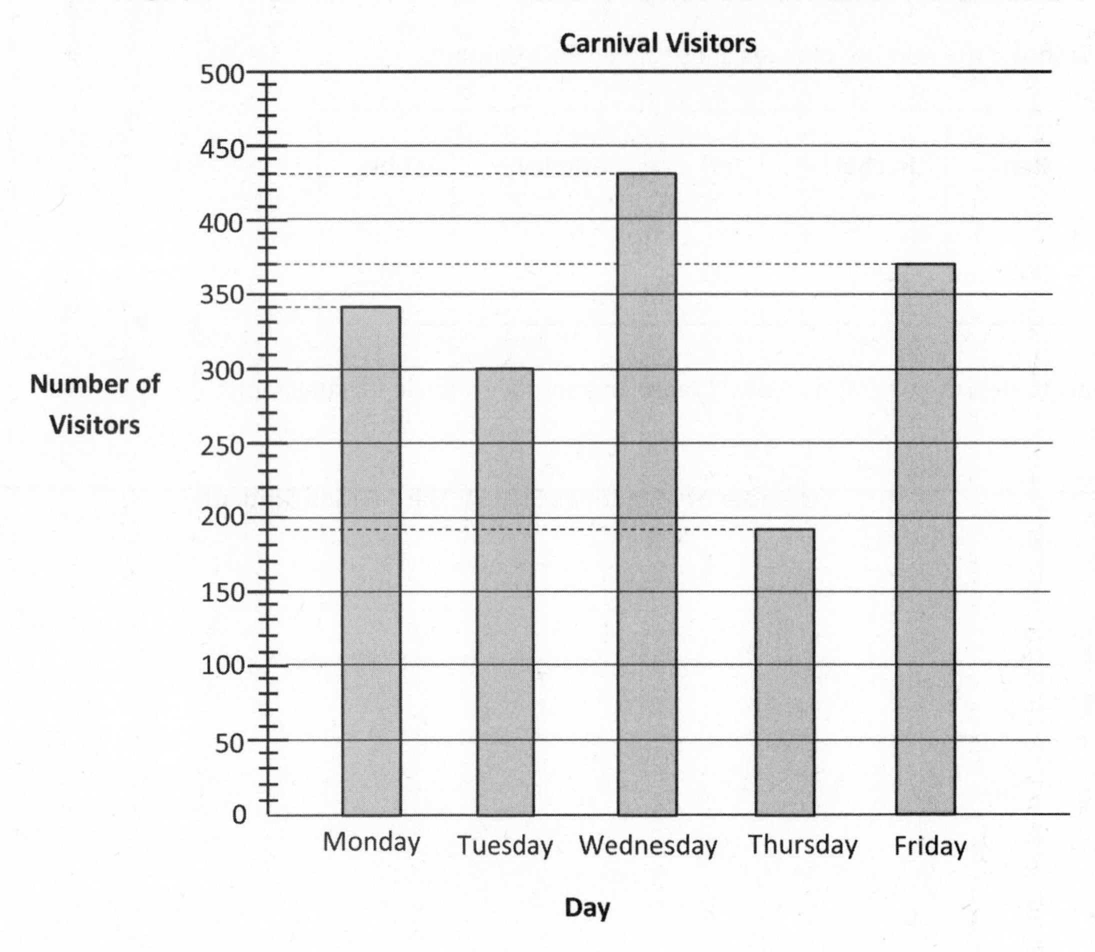

a. How many fewer visitors were there on the least busy day than on the busiest day?

b. How many more visitors attended the carnival on Monday and Tuesday combined than on Thursday and Friday combined?

Name ______________________________ Date ______________

The graph below shows the number of library books checked out in five days.

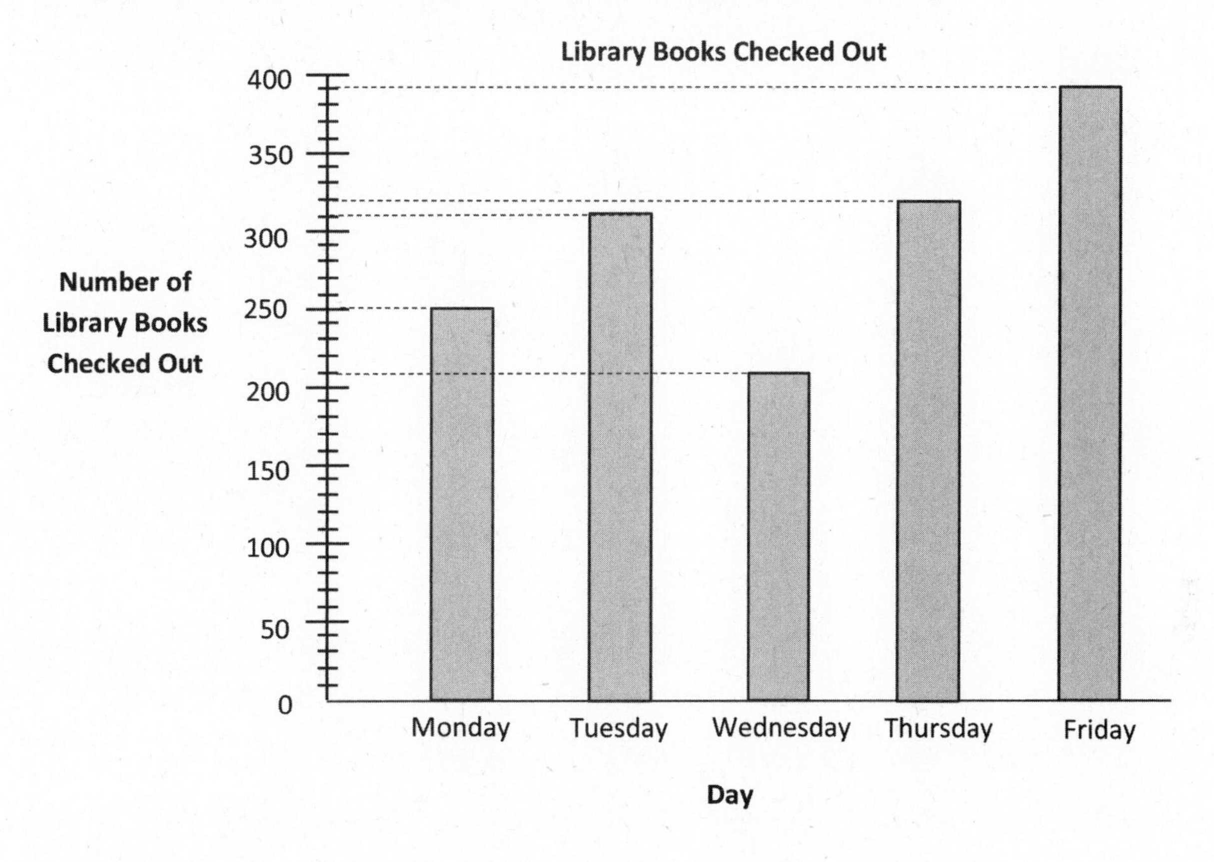

c. How many books in total were checked out on Wednesday and Thursday?

d. How many more books were checked out on Thursday and Friday than on Monday and Tuesday?

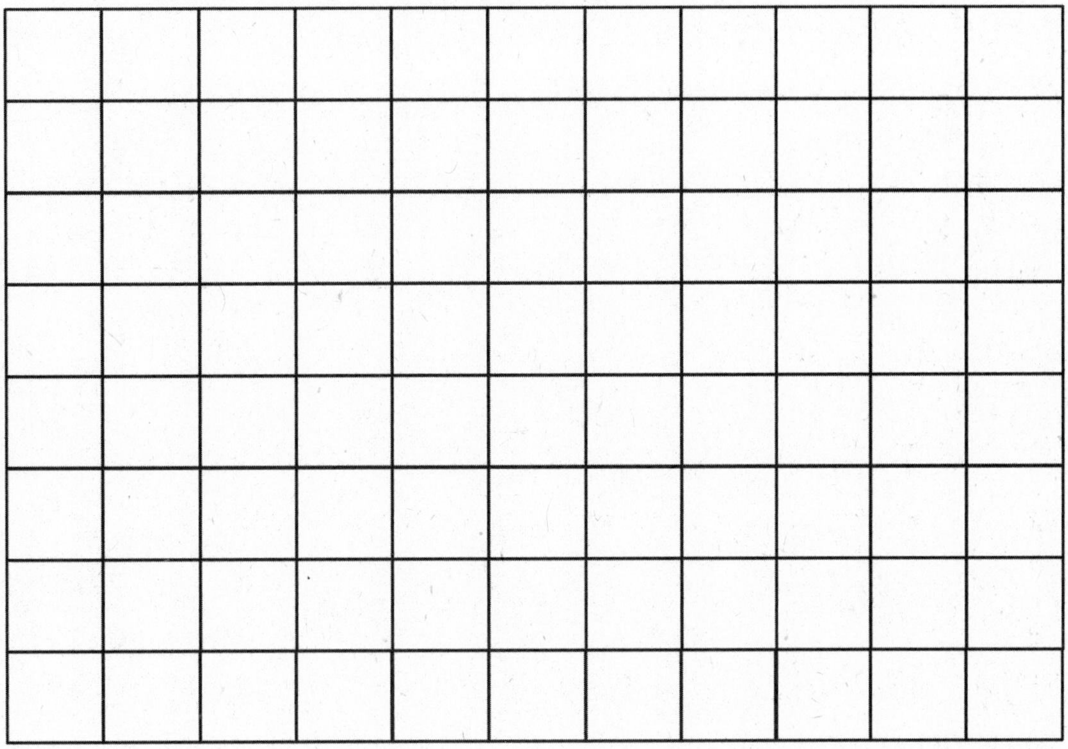

____

graph

Name ______________________________ Date ______________

1. Use the ruler you made to measure different classmates' straws to the nearest inch, $\frac{1}{2}$ inch, and $\frac{1}{4}$ inch. Record the measurements in the chart below. Draw a star next to measurements that are exact.

Straw Owner	Measured to the nearest inch	Measured to the nearest $\frac{1}{2}$ inch	Measured to the nearest $\frac{1}{4}$ inch
My straw			

a. ______________'s straw is the shortest straw I measured. It measures ______ inch(es).

b. ______________'s straw is the longest straw I measured. It measures ______ inches.

c. Choose the straw from your chart that was most accurately measured with the $\frac{1}{4}$-inch intervals on your ruler. How do you know the $\frac{1}{4}$-inch intervals are the most accurate for measuring this straw?

2. Jenna marks a 5-inch paper strip into equal parts as shown below.

a. Label the whole and half inches on the paper strip.

b. Estimate to draw the $\frac{1}{4}$-inch marks on the paper strip. Then, fill in the blanks below.

1 inch is equal to _______ half inches.

1 inch is equal to _______ quarter inches.

1 half inch is equal to _______ quarter inches.

c. Describe how Jenna could use this paper strip to measure an object that is longer than 5 inches.

3. Sari says her pencil measures 8 half inches. Bart disagrees and says it measures 4 inches. Explain to Bart why the two measurements are the same in the space below. Use words, pictures, or numbers.

Name ______________________________ Date ______________

Davon marks a 4-inch paper strip into equal parts as shown below.

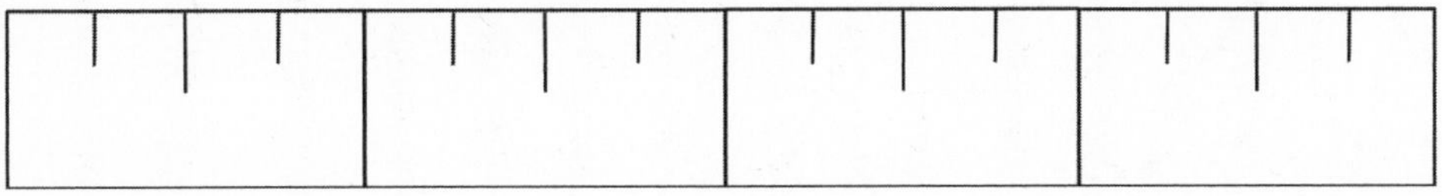

a. Label the whole and quarter inches on the paper strip.

b. Davon tells his teacher that his paper strip measures 4 inches. Sandra says it measures 16 quarter inches. Explain how the two measurements are the same. Use words, pictures, or numbers.

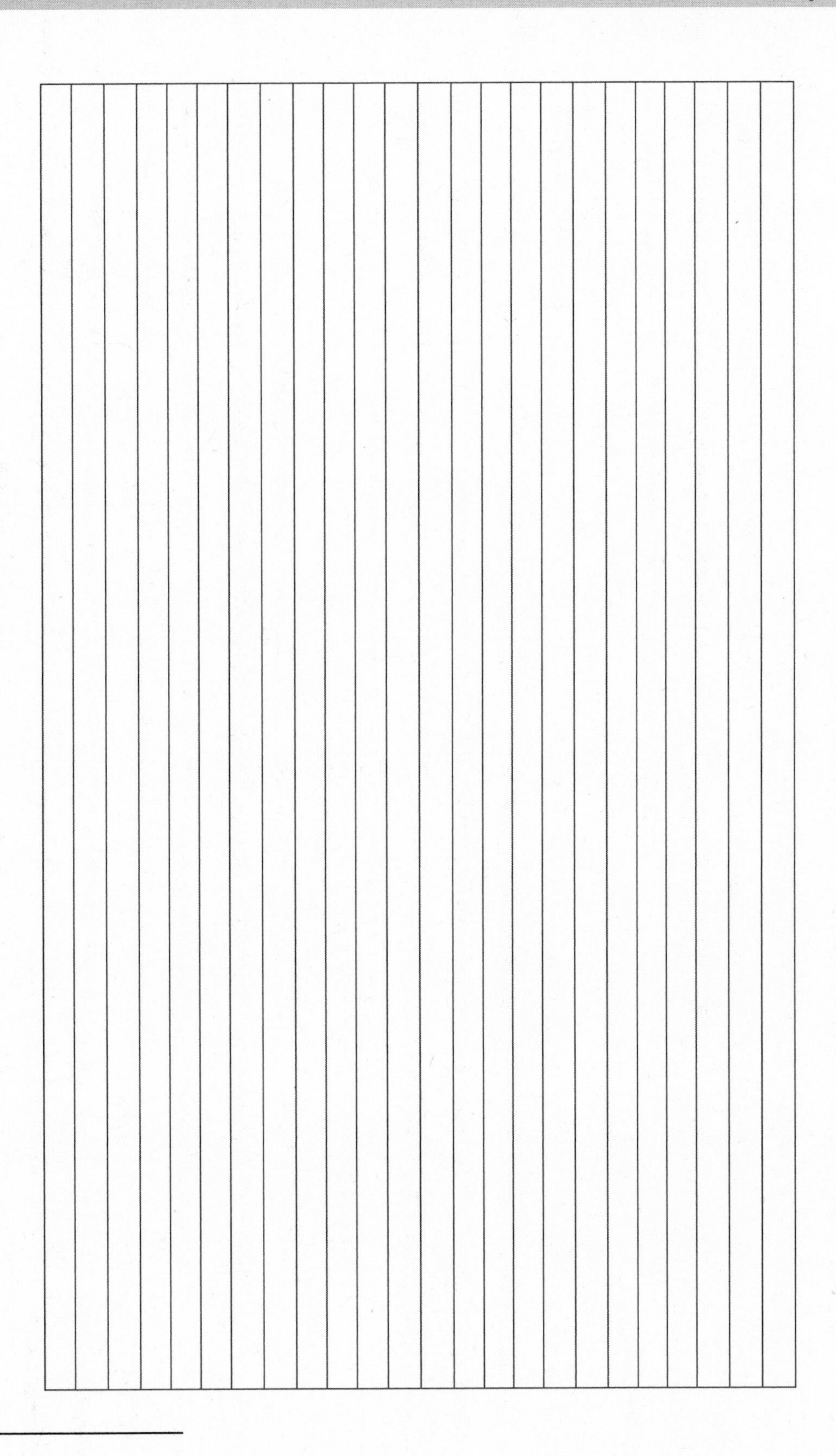

lined paper

Katelynn measures the height of her bean plant on Monday and again on Friday. She says that her bean plant grew 10 quarter inches. Her partner records $2\frac{1}{2}$ inches on his growth chart for the week. Is her partner right? Why or why not?

Read **Draw** **Write**

Name ______________________________ Date ______________

1. Coach Harris measures the heights of the children on his third-grade basketball team in inches. The heights are shown on the line plot below.

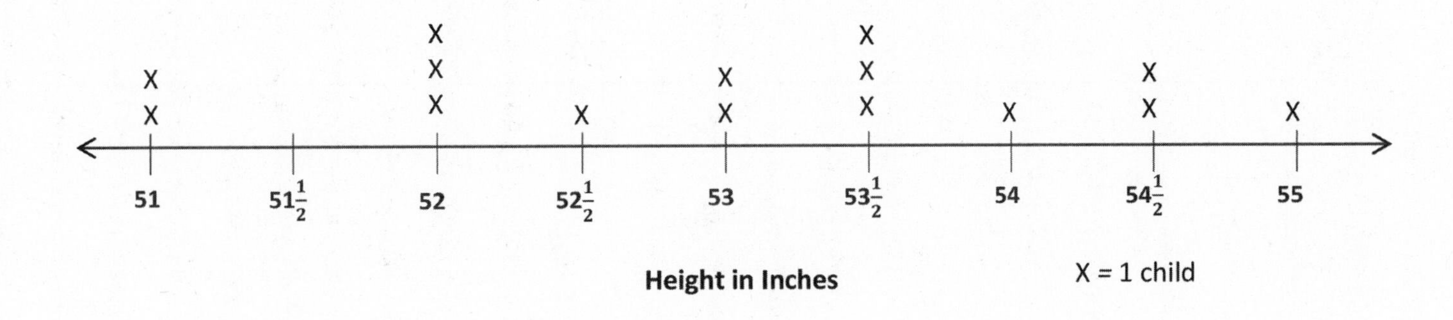

a. How many children are on the team? How do you know?

b. How many children are less than 53 inches tall?

c. Coach Harris says that the most common height for the children on his team is $53\frac{1}{2}$ inches. Is he right? Explain your answer.

d. Coach Harris says that the player who does the tip-off in the beginning of the game has to be at least 54 inches tall. How many children could do the tip-off?

2. Miss Vernier's class is studying worms. The lengths of the worms in inches are shown in the line plot below.

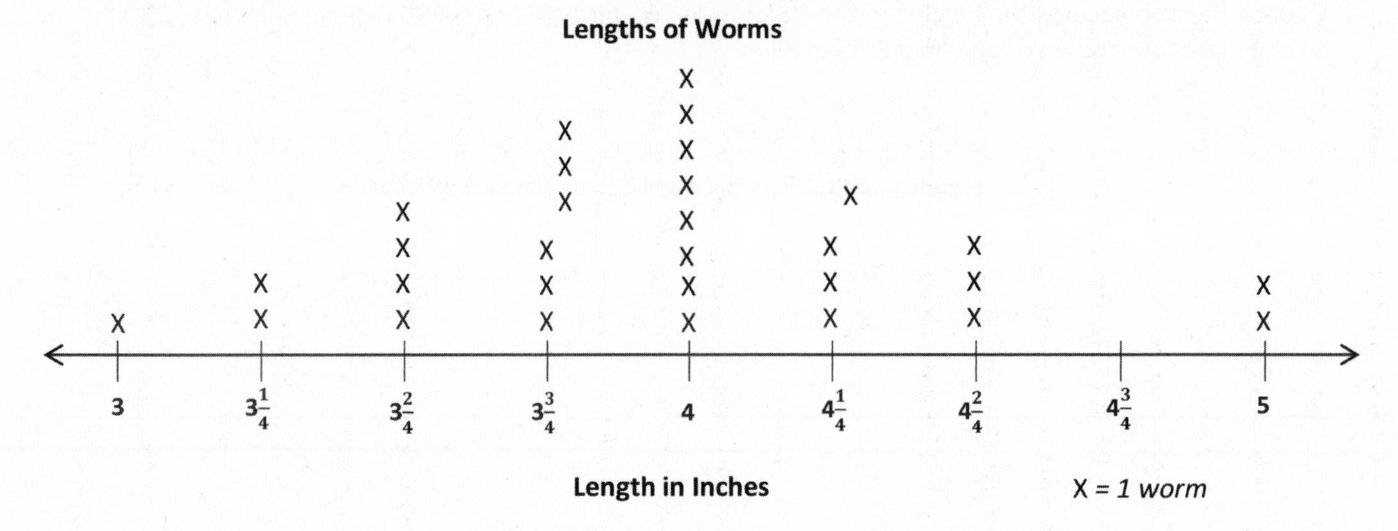

a. How many worms did the class measure? How do you know?

b. Cara says that there are more worms $3\frac{3}{4}$ inches long than worms that are $3\frac{2}{4}$ and $4\frac{1}{4}$ inches long combined. Is she right? Explain your answer.

c. Madeline finds a worm hiding under a leaf. She measures it, and it is $4\frac{3}{4}$ inches long. Plot the length of the worm on the line plot.

Name ______________________________ Date ____________

Ms. Bravo measures the lengths of her third-grade students' hands in inches. The lengths are shown on the line plot below.

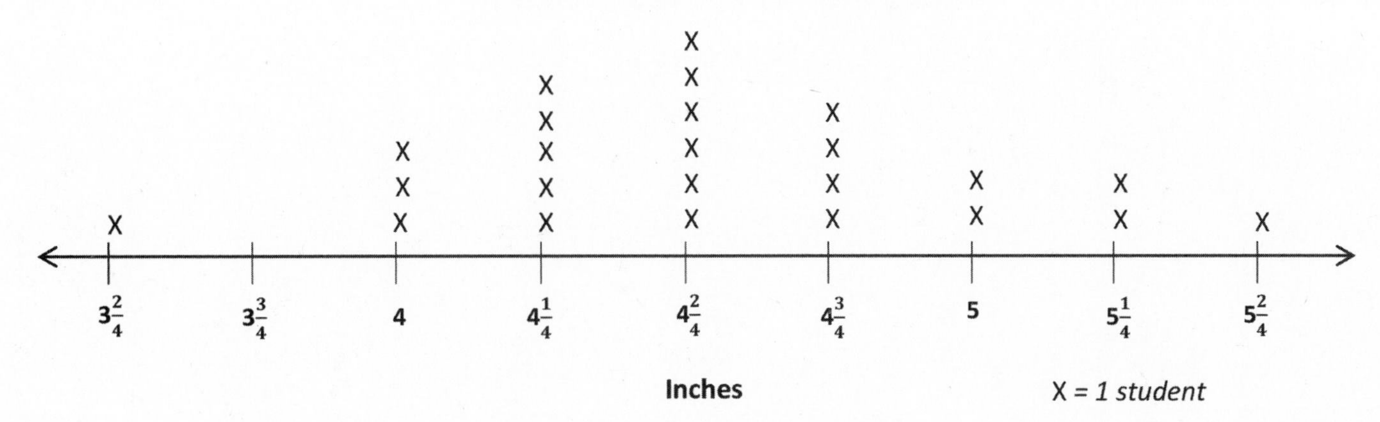

a. How many students are in Ms. Bravo's class? How do you know?

b. How many students' hands are longer than $4\frac{2}{4}$ inches?

c. Darren says that more students' hands are $4\frac{2}{4}$ inches long than 4 and $5\frac{1}{4}$ inches combined. Is he right? Explain your answer.

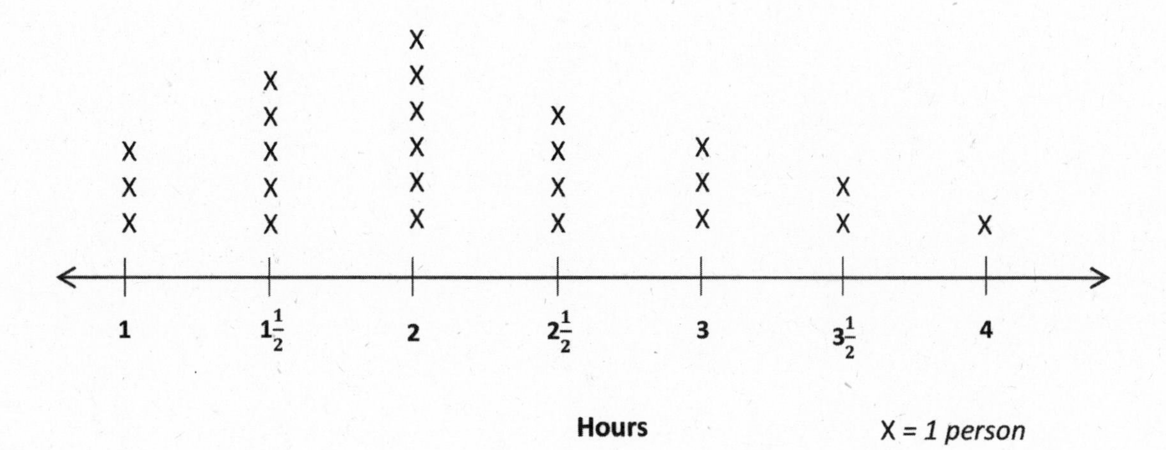

time spent outside line plot

The chart shows the lengths of straws measured in Mr. Han's class.

Straw Lengths (in Inches)				
3	4	$4\frac{1}{2}$	$2\frac{3}{4}$	$3\frac{3}{4}$
$3\frac{3}{4}$	$4\frac{1}{2}$	$3\frac{1}{4}$	4	$4\frac{3}{4}$
$4\frac{1}{4}$	5	3	$3\frac{1}{2}$	$4\frac{1}{2}$
$4\frac{1}{2}$	4	$3\frac{1}{4}$	5	$4\frac{1}{4}$

a. How many straws were measured? Explain how you know.

$4 \times 5 = 20$ straws

b. What is the smallest measurement on the chart? The greatest?

S - $2\frac{3}{4}$ in.

G - 5 in.

Read **Draw** **Write**

c. Were the straws measured to the nearest inch? How do you know?

No, because some have fractions.

Read **Draw** **Write**

Name ______________________ Date ____________

Mrs. Weisse's class grows beans for a science experiment. The students measure the heights of their bean plants to the nearest $\frac{1}{4}$ inch and record the measurements as shown below.

Heights of Bean Plants (in Inches)				
$2\frac{1}{4}$	$2\frac{3}{4}$	$3\frac{1}{4}$	$1\frac{3}{4}$	$1\frac{3}{4}$
$1\frac{3}{4}$	3	$2\frac{1}{2}$	$3\frac{1}{4}$	$2\frac{1}{2}$
2	$2\frac{1}{4}$	3	$2\frac{1}{4}$	3
$2\frac{1}{2}$	$3\frac{1}{4}$	$1\frac{3}{4}$	$2\frac{3}{4}$	2

a. Use the data to complete the line plot below.

Title: Heights of Bean Plants

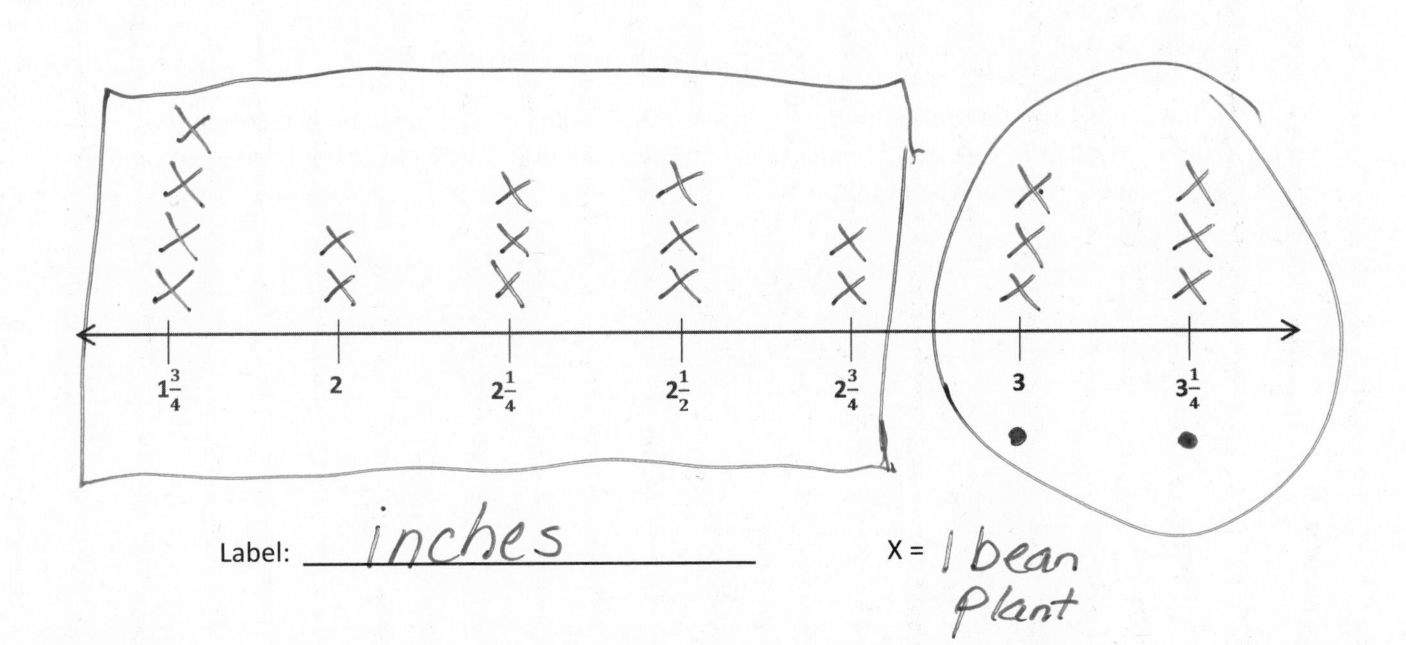

Label: inches X = 1 bean plant

b. How many bean plants are at least $2\frac{1}{4}$ inches tall?

14 plants

c. How many bean plants are taller than $2\frac{3}{4}$ inches?

6 plants

d. What is the most frequent measurement? How many bean plants were plotted for this measurement?

$1\frac{3}{4}$, 4 plants

e. George says that most of the bean plants are at least 3 inches tall. Is he right? Explain your answer.

George is incorrect. There are 6 plants that are at least 3 in. tall. There are 14 plants shorter than 3 in.

f. Savannah was absent the day the class measured the heights of their bean plants. When she returns, her plant measures $2\frac{2}{4}$ inches tall. Can Savannah plot the height of her bean plant on the class line plot? Why or why not?

Yes, because $2\frac{2}{4} = 2\frac{1}{2}$

Name ______________________________ Date ______________

Scientists measure the growth of mice in inches. The scientists measure the length of the mice to the nearest $\frac{1}{4}$ inch and record the measurements as shown below.

Lengths of Mice (in Inches)				
$3\frac{1}{4}$	3	$3\frac{1}{4}$	$3\frac{3}{4}$	4
$3\frac{3}{4}$	3	$4\frac{1}{2}$	$4\frac{1}{2}$	$3\frac{3}{4}$
4	$4\frac{1}{4}$	4	$4\frac{1}{4}$	4

Label each tick mark. Then, record the data on the line plot below.

Title: ______________________________

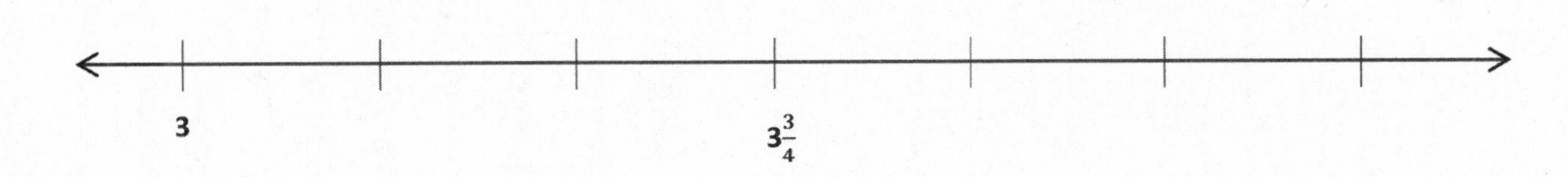

Label: ______________________________ X = 1 mouse

Straw Lengths (in Inches)				
3	4	$4\frac{1}{2}$	$2\frac{3}{4}$	$3\frac{3}{4}$
$3\frac{3}{4}$	$4\frac{1}{2}$	$3\frac{1}{4}$	4	$4\frac{3}{4}$
$4\frac{1}{4}$	5	3	$3\frac{1}{2}$	$4\frac{1}{2}$
$4\frac{3}{4}$	4	$3\frac{1}{4}$	5	$4\frac{1}{4}$

straw lengths

Mrs. Byrne's class is studying worms. They measure the lengths of the worms to the nearest quarter inch. The length of the shortest worm is $3\frac{3}{4}$ inches. The length of the longest worm is $5\frac{2}{4}$ inches. Kathleen says they need 8 quarter-inch intervals to plot the lengths of the worms on a line plot. Is she right? Why or why not?

Read Draw Write

Name ______________________________ Date ________________

Delilah stops under a silver maple tree and collects leaves. At home, she measures the widths of the leaves to the nearest $\frac{1}{4}$ inch and records the measurements as shown below.

Widths of Silver Maple Tree Leaves (in Inches)				
$5\frac{3}{4}$	6	$6\frac{1}{4}$	6	$5\frac{3}{4}$
$6\frac{1}{2}$	$6\frac{1}{4}$	$5\frac{1}{2}$	$5\frac{3}{4}$	6
$6\frac{1}{4}$	6	6	$6\frac{1}{2}$	$6\frac{1}{4}$
$6\frac{1}{2}$	$5\frac{3}{4}$	$6\frac{1}{4}$	6	$6\frac{3}{4}$
6	$6\frac{1}{4}$	6	$5\frac{3}{4}$	$6\frac{1}{2}$

a. Use the data to create a line plot below.

b. Explain the steps you took to create the line plot.

c. How many more leaves were 6 inches wide than $6\frac{1}{2}$ inches wide?

d. Find the three most frequent measurements on the line plot. What does this tell you about the typical width of a silver maple tree leaf?

Name ______________________________ Date ______________

The line plot below shows the lengths of fish the fishing boat caught.

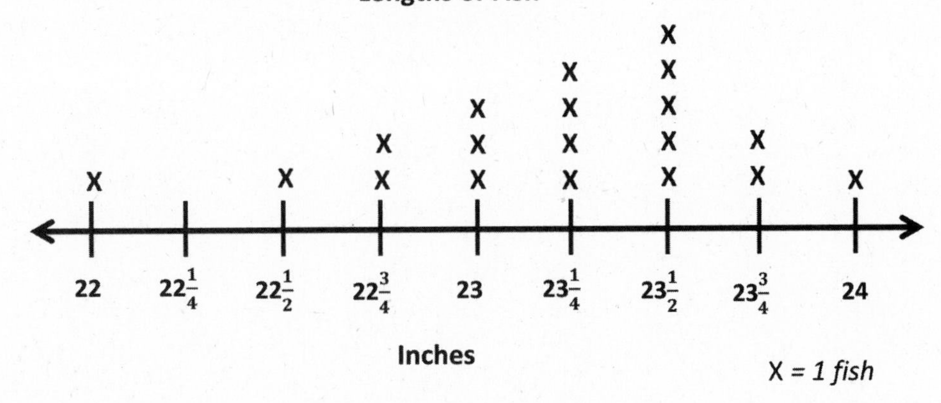

a. Find the three most frequent measurements on the line plot.

b. Find the difference between the lengths of the longest and shortest fish.

c. How many more fish were $23\frac{1}{4}$ inches long than 24 inches long?

Mrs. Schaut measures the heights of the sunflower plants in her garden. The measurements are shown in the chart below.

Heights of Sunflower Plants (in Inches)				
61	63	62	61	$62\frac{1}{2}$
$61\frac{1}{2}$	$61\frac{1}{2}$	$61\frac{1}{2}$	62	60
64	62	$60\frac{1}{2}$	$63\frac{1}{2}$	61
63	$62\frac{1}{2}$	$62\frac{1}{2}$	64	$62\frac{1}{2}$
$62\frac{1}{2}$	$63\frac{1}{2}$	63	$62\frac{1}{2}$	$63\frac{1}{2}$
62	$62\frac{1}{2}$	62	63	$60\frac{1}{2}$

heights of sunflower plants chart

Maria creates a line plot with a half-inch scale from 33 to 37 inches. How many tick marks should be on her line plot?

Read **Draw** **Write**

Name ______________________________ Date ______________

1. Four children went apple picking. The chart shows the number of apples the children picked.

Name	Number of Apples Picked
Stewart	16
Roxanne	____
Trisha	12
Philip	20
	Total: 72

a. Find the number of apples Roxanne picked to complete the chart.

b. Create a picture graph below using the data in the table.

Apples Picked

= ______ Apples

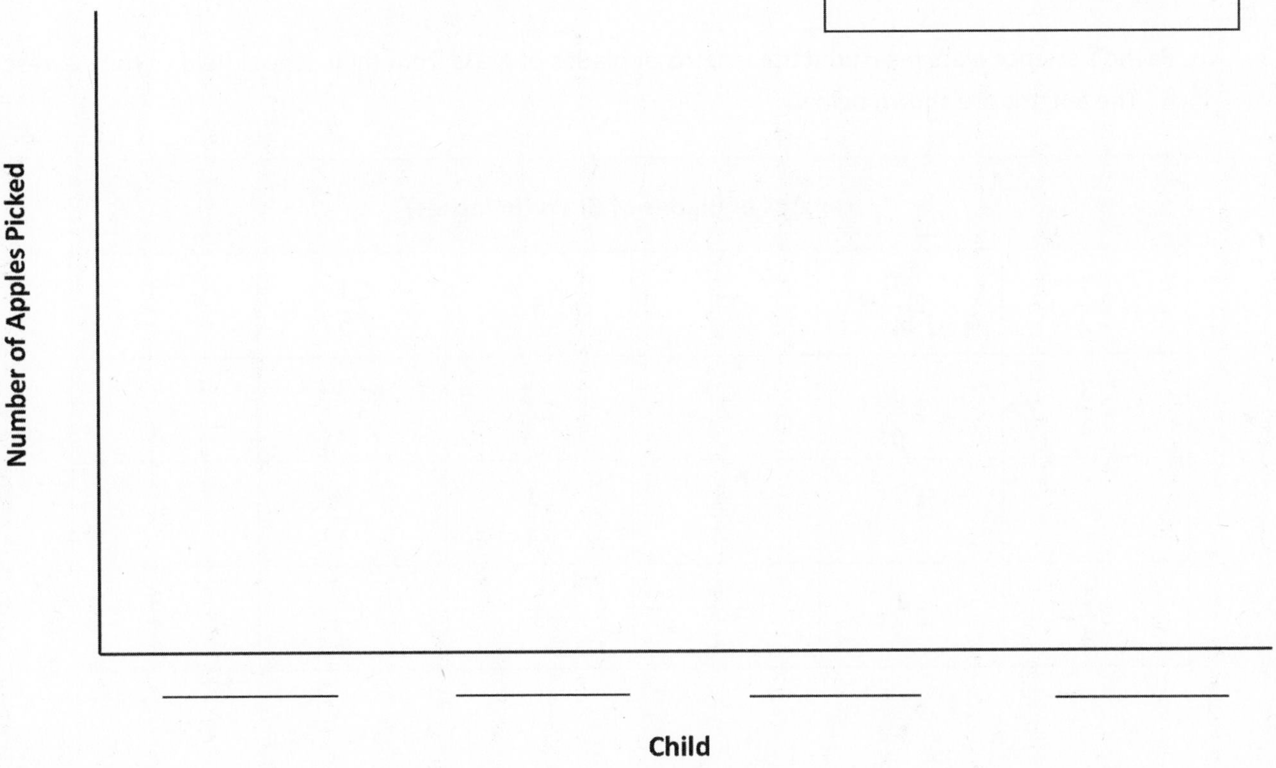

Number of Apples Picked

______ ______ ______ ______

Child

2. Use the chart or graph to answer the following questions.

   a. How many more apples did Stewart and Roxanne pick than Philip and Trisha?

   b. Trisha and Stewart combine their apples to make apples pies. Each pie takes 7 apples. How many pies can they make?

3. Ms. Pacho's science class measured the lengths of blades of grass from their school field to the nearest $\frac{1}{4}$ inch. The lengths are shown below.

Lengths of Blades of Grass (in Inches)					
$2\frac{1}{4}$	$2\frac{3}{4}$	$3\frac{1}{4}$	3	$2\frac{1}{2}$	$2\frac{3}{4}$
$2\frac{3}{4}$	$3\frac{3}{4}$	2	$2\frac{3}{4}$	$3\frac{3}{4}$	$3\frac{1}{4}$
3	$2\frac{1}{2}$	$3\frac{1}{4}$	$2\frac{1}{4}$	$2\frac{3}{4}$	3
$3\frac{1}{4}$	$2\frac{1}{4}$	$3\frac{3}{4}$	3	$3\frac{1}{4}$	$2\frac{3}{4}$

a. Make a line plot of the grass data. Explain your choice of scale.

b. How many blades of grass were measured? Explain how you know.

c. What was the length measured most frequently on the line plot? How many blades of grass had this length?

d. How many more blades of grass measured $2\frac{3}{4}$ inches than both $3\frac{3}{4}$ inches and 2 inches combined?

Name ______________________________ Date ________________

Mr. Gallagher's science class goes bird watching. The picture graph below shows the number of birds the class observes.

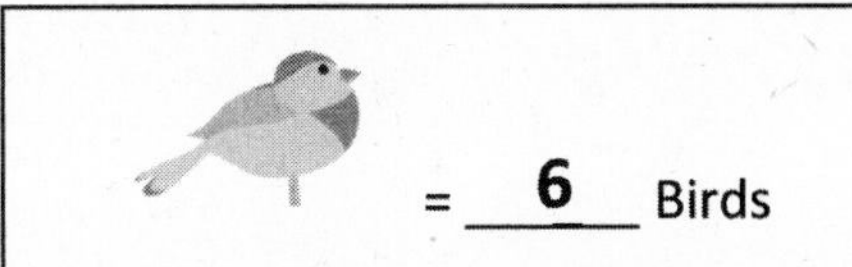

Number of Birds Mr. Gallagher's Class Observed

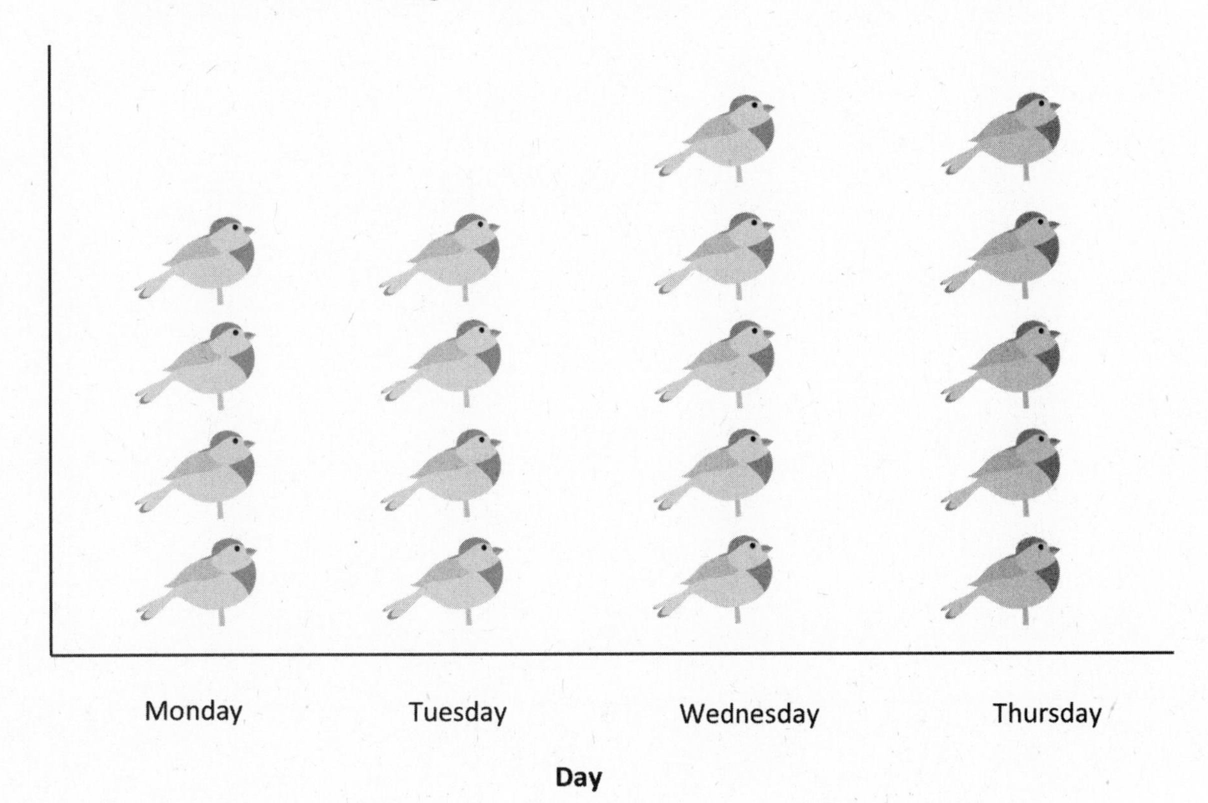

a. How many more birds did Mr. Gallagher's class observe on Wednesday and Thursday than Monday and Tuesday?

b. Mr. Manning's class observed 104 birds. How many more birds did Mr. Gallagher's class observe?

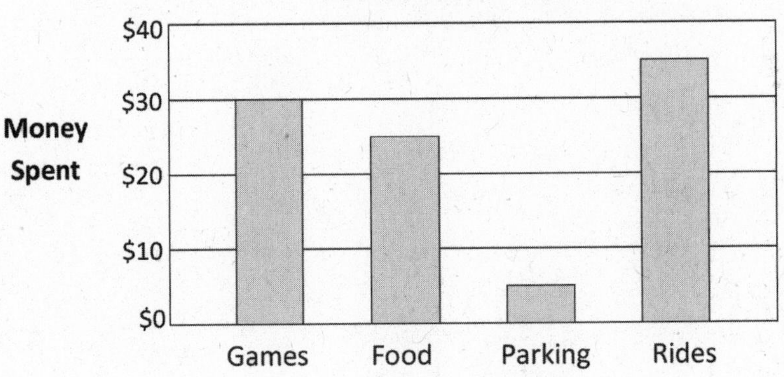

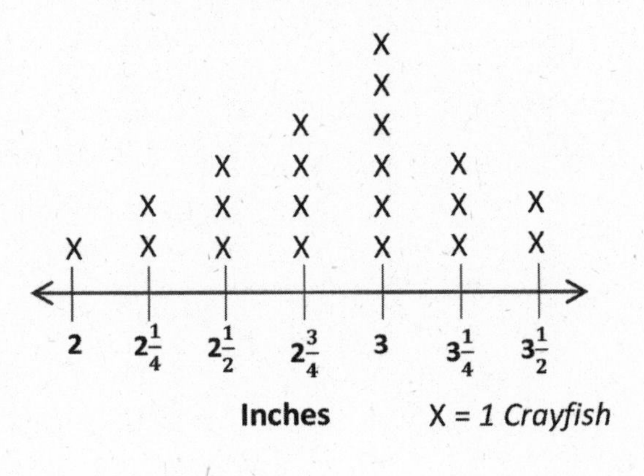

bar graph and line plot

Credits

Great Minds® has made every effort to obtain permission for the reprinting of all copyrighted material. If any owner of copyrighted material is not acknowledged herein, please contact Great Minds for proper acknowledgment in all future editions and reprints of this module.